AF541580

Insect Pollinators

NIPA® GENX ELECTRONIC RESOURCES & SOLUTIONS P. LTD.
New Delhi-110 034

About the Author

Dr. Akhilesh Kumar Singh, is working as Associate Professor, Department of Plant Protection, College of Horticulture, Banda University of Agriculture and Technology Banda -210001 (U.P.) India. He has served as Assistant Scientist about 7 years and currently serving in the teaching profession for more than 6 years. He did Bachelor's degree in Agriculture in the year 1999, Masters in Entomology in 2002 from the prestigious ANDUAT, Ayodhya and accomplished Doctoral Research in Entomology and Agricultural Zoology and conferred Ph. D. from BHU, Varanasi in 2006. He is also qualified ASRB (NET) in the field of Entomology. He has accomplished 12 externally funded Projects as Principal Investigator of worth Rs. 4730.63 Lakh and currently serving in 3 externally funded research projects as Principal Investigator.

Besides contributing more than 72 publications, which include 42 research papers in peer reviewed journals, 1 book and 6 book chapters, 23 articles, etc. He has organized 2 seminars, attended 6 international, 13 national seminars and presented papers and also delivered 18 invited lectures in advance training programs, farmers training programme.

Dr. Singh guided two M.Sc. Students and being a member of evaluation committee of six M.Sc. Students. He conferred several awards viz. Agri-innovation Award in 2015 from GKV Society, Agra, UP, Young Scientist Award 2016 from S&T SIRI, Telangan, Excellence in Research Award (Senior Category) in 2021 from Banda University of Agriculture and Technology Banda, Banda, Best paper Award in 2020 from Society for horticulture research & development, Best Paper Award in 2013 from Society for advancement of natural resins and gums, IINRG Ranchi. He is Fellow of Entomological Society of India and Indian Ecological Society and member of different National and international societies.

He is always willing to provide his services to the farming community, social service, educational society and to help the students.

Insect Pollinators

Akhilesh Kumar Singh
Associate Professor
Department of Plant Protection
College of Horticulture
Banda University of Agriculture and Technology
Banda -210001 (U.P.) India

NIPA® GENX ELECTRONIC RESOURCES & SOLUTIONS P. LTD.
New Delhi-110 034

NIPA® GENX ELECTRONIC RESOURCES & SOLUTIONS P. LTD.

101,103, Vikas Surya Plaza, CU Block
L.S.C.Market, Pitam Pura, New Delhi-110 034
Ph : +91 11 27341616, 27341717, 27341718
E-mail:newindiapublishingagency@gmail.com
www: www.nipabooks.com

For customer assistance, please contact
Phone: + 91-11-27 34 17 17
Fax: + 91-11- 27 34 16 16

ISBN: 978-81-19235-66-7

Composed and Designed by NIPA®.

Acknowledgement

I am grateful to AICRP (Honey Bees & Pollinators) for financial assistance supported by ICAR, New Delhi, India, and the host institution Nagaland University.

I am thankful to my colleagues; Mr. Ngukho and Mr. Akum, for their help and support. I would like to thank Dr. H. K. Singh for their valuable support.

A.K. Singh

Vice Chancellor Office

Banda University of Agriculture & Technology, Banda-210001, U.P.
Tele Fax No:- 05192-232305Ph-05192-232305;
website: buat.edu.in; email-vc.buat@gmail.com, vcbuat.bn-up@gov.in

Ref.:- VC/2023/261 Date: 08/11/2023

Dr. Narendra Pratap Singh
Vice Chancellor

Message

Pollination is an essential ecosystem service that enables plant reproduction and food production. Population has been proliferating day by day. Indian population has reached about 125 crore. Food requirement have positive correlation with population and therefore, food productivity needs to increase day by day for sustenance. The Govt. of India already passed the bill of food security in 2014. For the population explosion, ensuring food security requires enhancing the productivity as more food which may be fulfilled by land availability is also limited. Food security, food diversity, human nutrition and plant diversity all rely strongly on pollinators and mostly on insect pollinators. Augmentation of pollinators and sufficient pollination is a tool to enhance the productivity.

Although more then 750,000 insect pollinators have been described, only about 16,000 species of bees are known in the world and only a few species have been managed for crop pollination. These include honey bees, bumble bees, leaf-cutting bee, mason bees, alkali bees, carpenter bees and stingless bees. Agriculture biodiversity is often understood as crop genetic resources, yet agro-ecosystems hold a wide diversity of other organisms that contribute toward their productivity and sustainability.

The book on "Insect Pollinator" contains a comprehensive report about insect pollinators, constraints, and their management. The effort put forth in preparation of this book by Dr. A. K. Singh (Associate Professor) Department of Plant Protection, Banda University of Agriculture & Technology, Banda, deserve appreciation and I congratulate him to bring out this informative publication.

(Narendra Pratap Singh)
Vice Chancellor

उद्यान महाविद्यालय
बांदा कृषि एवं प्रौद्योगिक विश्वविद्यालय बांदा-210001, उ०प्र०
Banda University of Agriculture & Technology
Banda-210001, U.P.
Tele No:- 05192- 232303; fax: 05192-232305;
Website:- buat.edu.in, e-mail:-com ptroller. mskjuat@qmail. com

Dr. S.V. Dwivedi
Dean College of Horticulture

Message

Pollinators provide an essential ecosystem service, namely pollination and it has an intricate relationship between plants and animals; reduction or loss of either will affect the survival of both. Pollinator's bio-resource is an endowment for pollination and production of valuable commodities for human/ animal life. It has been estimated that 80 per cent of plants pollination depends upon pollinators and mostly by insects. In agriculture, enhancement of crop productivity has seen a major ascent by mobilizing such inputs that has free or low cost available in nature. By manipulating agro-ecologies and agro techniques service of pollinator fauna could be enhanced the productivity. The current agricultural practices through large scale monoculture, pest management dominated by pesticides, destruction of natural habitats and intensive tillage have caused decline in the abundance and diversity of crop insect pollinators. Conservation and augmentation of pollinators can proliferate the pollinators for fulfillment of pollination deficits. In agro-ecosystems, management of pollinators and pollination input can be maximized for secured crop productivity and enhancement of the yield quality.

I appreciate the effort made by Dr. A. K. Singh, Associate Professor, Department of Plant Protection, Banda University of Agriculture & Technology, Banda, for documenting and publication of Book on Insect Pollinators. I am sure the comprehensive information of insect pollinators will enhance the knowledge of researchers, students, extension workers and farmers.

(S.V. Dwivedi)
Dean
College of Horticulture

Banaras Hindu University
Varanasi, Uttar Pradesh-221005

Dr. S.V.S. Raju
Director IAS, BHU

Foreword

The importance of pollination in agro-ecosystem and forest ecosystem has been recognized for millennia. About 80 per cent of flowering plant species are dedicated on pollination by animals, mostly insects. However, pollinators population have declined and some extinction have been observed. The erosion of pollinator population and diversity resulting in pollination crisis significantly affects the crop productivity, food security, plant diversity, wider ecosystem stability and human welfare. Pollinators' abundance and diversity erosion is due to mankind activities viz. destruction of natural habitats, intensive tillage, monoculture, injudicious use of pesticides, climate change and pests. In modern agro-ecosystem, highly cross-pollinated commercial crops require pollinators' augmentation for maximization of their productivity. Insufficient pollination may cause due to lack of pollinators and may shows pollination crisis. Pollination depends, to a large extent, on the symbiosis between plant species and their pollinators which results into intricate relationships between plant and animal - the reduction or loss of either affecting the survival of both. Pollinators are an input of crop cultivation, associated biodiversity and provide an essential ecosystem service to both natural and agro- ecosystems. In agroecosystems, management of pollinators and pollination input can be utilized to maximize crop productivity and enhance the yield quality. To be provident by virtue of pollinators' abundance and diversity, we should conserve and as far as possible, proliferate the population till asymptote is achieved for fulfilment of required pollination of plant kingdom. I am happy to learn that efforts have been made to provide information about insect pollinators and their management in the form of a book "Insect Pollinators". The sacrificial efforts put forth in preparation of this book by the author, Dr. A. K. Singh, Associate Professor, Department of Plant Protection, Banda University of Agriculture & Technology, Banda bringing out this timely publication deserves appreciation.

(S.V.S. Raju)
Director IAS, BHU

Preface

Pollination is one of the crucial ecosystem services. About 80% of the angiosperm species are dedicated on animal pollination, mostly by insect. In agricultural ecosystem, many agricultural crops are dependent on insects for their pollination, and assisted pollination may have to be done when natural pollination is insufficient in order to reduce potential yield loss. The economic value of the pollination services provided by insect pollinators in 2005 was estimated to be about €153 billion. The United Nations also reported on the economics of ecosystems and biodiversity that insect pollination was valued at £134 billion. There is the irony to ignore the precious insect pollinators and their service which is a key constraint to words the sustainability of contemporary agricultural practices.

Pollinators belong to diverse groups of the animal kingdom, including birds, bat, reptiles, insects, etc. Many plants have evolved intricate relationships with many insect pollinators, without which majority of plant species would not reproduce and maintain their genetic diversity. Conservation of pollinator's biodiversity is important for the potential yield of agricultural as well as horticultural crops and their hybrid seed production. Insect pollinators and other pollinators are declining at an alarming rate which has threatened the existence of plant life and this downward trend could damage dozens of commercially important crops. The decline in pollinators' population and diversity is a serious concern for agricultural production, conservation and maintenance of biodiversity in many parts of the world.

Conservation of pollinator's biodiversity is important for the potential yield of agricultural as well as horticultural crops and their hybrid seed production. Encouragement of wild pollinators, domestication of unused potential pollinators and more environmentally, sensitive human exploitation of the world are needed as part of conservation, forestry, agro forestry, sustainable agriculture and development. For providence it is our need to conserve and as much as possible, proliferate the pollinators' population till asymptote crosses the boundaries between agrarianism and land ethic, despite the fact that the values and goals of farmers and biodiversity conservationists are not always in synchrony with each other.

Contents

1

Insect Pollinators

Pollination is one of the most essential ecosystem services which is effectual for crop productivity, biodiversity and human livelihood. Pollination is the process of transferring pollens from anther to stigma for plant reproduction and has direct impact on crop productivity. Conservation and management of pollinators is still ignored in the agriculture development program by the policy makers. In agricultural ecosystem, many agricultural crops are dependent on insects for their pollination, and assisted pollination may have to be done when natural pollination is insufficient in order to reduce potential yield loss (Klein *et al*., 2007). About 80% of the world's flowering plant species are dedicated to animal pollination, mostly by insects (FAO, 2007). In 2005, globally, pollinators' pollination services economic value was estimated to be about €153 billion (Gallai *et al.,* 2009). The United Nations also reported that the economics of ecosystems and biodiversity by insect pollination was valued at £134 billion.

There is the irony to ignore the precious insect pollinators and their service which is a key constraint to the sustainability of contemporary agricultural practices. It involves the process of securing pollinators effectively towards agricultural pollination through the conservation and augmentation support of pollinators. Pollinators belong to diverse groups of the animalia kingdom, including birds, bats, reptiles, insects, etc. Conservation of pollinator's biodiversity is important for the potential yield of agricultural as well as horticultural crops and their hybrid seed production. Encouragement of pollinators, domestication of unexploited potential pollinators are needed as part of conservation, forestry, agro-forestry, sustainable agriculture and development. For improved awareness among mankind community, vibrant programs are required to urge the policy-makers, researchers, extension workers, farmers, etc. It may influence pollinator conservation and sustainable use towards congenial practices for conservation of pollinators.

Bees are dominant pollinators in tropical region because their food sources are totally dependent upon flowers (Roubic, 1989; Michner, 2007). The extent of our reliance on single species for such an important service is risky. Therefore, an additional role can be played by native and wild pollinators, provided that attention is given to curtailing of population caused by inadvertent insecticide poisoning, habitat destruction, climate change and monoculture. About 75% of genetically bio-diverse agricultural crops have been lost since the beginning of 20th century from the earth and 25% of the world's plant species were endangered in 1980 and which will be extinct by 2015 (FAO, 1993). An estimated 62% of the flowering plants may be suffering with reduced regeneration from seeds as a result of pollinator scarcity (Burd, 1994). More diverse communities of pollinators in agricultural systems are more beneficial for crop pollination (Pritchard, 2005 and Chacoff and Aizen, 2006).

Pollination: It is the process by which pollen is transferred from the anther to the stigma of the same species of angiosperms' flower, thereby enabling fertilization and reproduction.

Angiosperm: It is the flower bearing and seed-producing plants which are the most diverse group of land plants. It is also known as Angiospermae or Magnoliophyta.

Pollenizer: It is a plant that is a source of pollen for the pollination process.

Pollinator: It is the biotic agent that transfers pollen from the anthers of a flower to the stigma of a flower to accomplish fertilization or 'syngamy' of the female gamete in the ovule of the flower by the male gamete from the pollen grain. If insects play this role as a biotic agent it is known as insect pollinator.

The Plant Pollination Process

The following steps are accomplished during pollination process,

1. Dehiscence of anther,
2. Pollen grains transfer on the receptive stigma,
3. A pollen tube grows down the style,
4. The sperm nuclei fuse with the female ovules,
5. The ovules develop as seed, and the ovary develops as fruit,

Abiotic Pollination

About 20 per cent of plant species depend upon abiotic pollination. Some plants, especially grasses, mostly conifers and some deciduous trees are pollinated by wind and this is called 'anemophily'. The structure of the plant is adapted to enable pollen grains to be blown from one plant to another. Anemophily plant's anthers usually produce huge number of pollens and these pollens are which are dry and light weight. Abiotic pollination is also divided on the basis of pollination mode; Gravity pollination (Geophilly), wind pollination (Anemophily) and water pollination (Hydrophily).

Biotic Pollination (Zoophily)

About 80 per cent plant species pollination requires the help of animals such as insects, birds (humming bird), bats, reptiles, etc. to transfer pollen from one flower to another flower of same species and this is called biotic pollination. The major pollinators belong to Hymenoptera order and out of this most belong to Apidae family. Other animals such as bats, birds, butterflies, moths, flies and beetles also play key roles in pollination. Insect pollination is also known as 'entomophily'.

Insect Pollinator

Insect pollinated plant species usually have dry pollen while stigma is viscid. When any pollinator touches the latter some mucous is transferred on its body and thereafter, when this pollinator visits another flower, the adhered loose pollen on its body is deposited on the stigma of that flower. Although more than 750,000 insect pollinators have been described (Grimaldi and Engel, 2005), possibly as many as 30 million more await discovery and formal description (Erwin, 2004). Fortunately, about 16,000 species of bees are known in the world (Michener, 2000); only a few species, however, are managed specifically as crop pollinators.

The majority of angiosperms depend on animal pollination through insects, birds, bats, reptiles, acaroids and others. Since insects play the undisputed pivotal role in pollination, their efficiency towards pollination should be utilized to the maximum because they have immense diversity with great co-evolution ability in Animalia Kingdome. Pollination endow is a key ecosystem service for food security. Pollinators are essential for many fruit and vegetable crops. In agriculture, especially amongst cross pollinated and pollen- scanty crops, essentially require true pollinator for Eu- pollination services to enhance the crop productivity and minimize the extravagant agricultural inputs. Many plants have evolved intricate relationships with many insect pollinators, without which they would not reproduce and maintain their genetic diversity (Daily *et al.,*1997). Most of the world's staple foods including wheat, corn and rice reproduce without insect pollination. These crops account for 65 per cent of global food production, still leaving as much as 35 per cent depending on pollinating animals (Klein *et al.,* 2007). In part, due to the massive scale and homogeneity of modern agriculture, the majority of crops requiring pollination are dependent on managed pollinators, and especially on managed honey bees (Aizen *et al.*, 2008).

Pollination is a natural endowment and so we tend to take it for granted, and that speculation is our problem. It is a fact that 80 per cent of flowering plant species rely on insect pollinators (FA0, 2007). The global value of the pollination service has been recently estimated to be about $200 billion (Gallai *et al.*, 2009). In agro-ecosystem, many crops depend upon insects for their pollination, and according to the requirement, assisted pollination may have to be done when natural pollination is insufficient in order to reduce potential yield loss (Klein *et al.*, 2007). In natural ecosystem, insect pollination plays a pivotal role in maintaining and conserving biodiversity. The United Nations reported that the economics of ecosystems and biodiversity by insect pollination was valued at £134 billion.

Bees are the most important pollinators of many crops and are recorded as visitors to 73 per cent of the crop species (Nabhan and Buchmann, 1997). Honey bees belong to the genus *Apis* of the family Apidae, order Hymenoptera. Amongst insects, bees are ultimate pollinators, especially in cross pollinated crops. The other insect pollinators go into hibernation at certain times in a year while bees are active the whole year for pollination; the working period (foraging duration) of honey bee is longer in a day than other insects and the number of bees per colony is also higher than other insects. All these attributes rank bees as the more effective pollinators among insects. Honey bees (*Apis*) alone pollinate crops that have an added value of over $14 billion (Abrol, 2012).

Among the insect pollinators, solitary and social bees provide most pollination in both managed and natural ecosystems. There was a strong linear relationship however, between per-visit honey bee pollination efficiency and the richness and abundance of wild bees present, increasing the number of seeds set per visit thereby deposition of pollen on stigma. During foraging internal competition observe between wild bees and honey bees where usually honey bees transfer pollen more frequently with more number of pollen deposition from anther to stigma and enhance their per-visit efficiency. The better seed and fruit set can result from the combined, complementary foraging activities of honey bees plus wild bees than either group alone. In some crops (alfalfa, blueberry, and cranberry) Non-*Apis* bees are more effective pollinators than *Apis mellifera* for productivity.

Various orders of class Insecta play vital role in the pollination of angiosperm and are described below:

Hymenoptera

Bee Pollination (Melitophily)

Bee belongs to the order hymenoptera; family Apidae. The Apidae family is a largest family of super family Apoidea containing about 5700 species of bees. They live either socially or as solitary insects. The body is covered with dense hairs and have chewing and lapping type mouthparts. They collect nectar or pollen or both and store in hive. Among the insect pollinators, hymenoptera order is a key order which have the largest and diversified assemblage of beneficial insects; whereas Buchman and Ascher (2005) affirmed that bee species pollinate about 250,000 described angiosperms. Hymenoptera order contains 25,000 to 30,000 species including bees, wasps, ants, chalcids, ichneumons, sawflies, etc. belonging to 250 genera and 11 families regarding the most important pollinators. The pollinators of Hymenoptera order may be divided into two parts Apis pollinator and Non-Apis pollinator. Their morphology and behaviours make them effectual and paramount pollinator for a wide range of angiosperms.

Eu-characters ofTrue Pollinator in Honey bees

Honey bees have gradually evolved numerous Eu-characters of true pollinators which make them the most proficient and effectual pollinators.

1 Honey bee species are social insects and all the cast live together in a colony. In the colony, worker bee population is in thousands where three weeks old workers attend to outside duties like food collection and pollination. The forager's population are also in thousands, which is much higher than other pollinators in a single colony except stingless bee. More workers must attend to more number of flowers and more frequent visitation.

2. Most of the body part of the honey bee are covered with dense and forked hairs and these hairs are more capable to hold the pollens as compared to simple and un-forked hairs.

3. Worker honey bees have two corbiculae (pollen basket) on the hind legs of the worker bee which allow it to carry pollen for their consumption. These corbicula pollen do not use in pollination. Only loose pollen which are adhered on other body parts are used in pollination.

4. They have specialized chewing and lapping type mouth; they have long labium which is reachable to the nectarine of flowers and well developed honey sac to store the nectar during foraging. Searching nectar on flowers maximise visit on flower and thereby maximise pollination.
5. They have ingenious communication system among their colony colleague through dances and other specialized systems of communication; scout bees can transmit this information to other members of the colony, thus greatly improving pollination activity.
6. They do not go for hibernation unlike other insects and continue pollination process the whole year although their foraging may discontinue during extreme weather conditions.
7. Their precise flower handling ability is commendable; they meticulously accomplish pollination during foraging without causing any harm to flowers which is incredible.
8. Honey bee foraging commences early and cessation is much delayed than other pollinators and therefore, its working period is longer *viz.* "first come- last go".
9. Each colony contains a large number of worker bees for foraging within a narrow pollinating window of particular crop. Bees will usually forage on only one species of flower on each trip and due to this floral constancy, honey bees are desirable pollinator for most of the crop for seed production as well as to enhance the productivity.
10. High foraging rate, foraging speed and minimum transit time makes they ultimate pollinators.

Non-apis Pollinators

The pollination deficit is increasing with depletion of pollinator population and their extinction and therefore, the need of pollination services will greatly increase for food security to the increasing population in the near future. *Apis mellifera* dominated commercial pollination practice is famous globally due to their Eu-social behaviours and true pollinator attributes. Our extent of reliability on a single pollinator for such a precious service is much risky. The wild bees are also precious pollinators and their contribution in pollination is underestimated due to their low population abundance, foraging behaviour, nesting behaviour and lack of knowledge. The other reason may be that we rely more on the easily manageable honey bees which provide bee hive products also. Honey bee population depletion has created a pressure on scientific community and especially, the declining *Apis mellifera* population due to CCD (Colony Collapse Disorder). The non-Apis bees effectively complement honey bee pollination in many crops *viz.* bumble bees for solenaceous crops which require buzz pollination, solitary bees Nomia and Osmia for the pollination of orchard crops, Megachile for alfalfa pollination, and social stingless bees to pollinate coffee and other crops.

There are about 19000 described species of bees in the world (Linsley, 1958) and, usually entire bees are categorized in the group of wild bee except *Apis mellifera*. The term "wild bee" is used commonly for all bees except honey bees in the genus Apis. According to an estimate, there are at least 30,000 species of bees in the world (Abrol, 2012). The nesting behaviour of most non-apis species are either single or complex nests underground and constructed with earthen, leaf or resin nests on rocks and plants whereas some adopt to crevices in rocks or plant stems, abandoned nest of mammals and plant galls for their nesting sites. Most of these bees live a solitary existence and each female after mating, locates and builds her nest without the aid of other bees.

Short-tongued Bees

The Andrenidae, Colletidae, Halictidae, Mellittidae and Oxaedae are characterized by usually short mouthparts and are often grouped as the short-tongued bees. The other four families Anthophoridae, Megachilidae and Apidae are classified as long tongued families. Usually, their mouthpart modifications observed are elongated glossae, maxillary galeae and basal segments of the labial palpi forming the sucking apparatus for taking advantage of nectarines from deep tubular flowers. Flowers with deep corolla tubes probably arose in co-evolution with their principal pollinators. The legume crops and their pollinators such as Megachilidea bees constitute an important example.

Wasp Pollination (Sphecophily)

Usually, their foraging behaviours are unsteady and so they are considered as a partial pollinator. They have primitive mouthparts with small and flat tongue which are used for lapping nectar. Their hard and rude bodies covered sparsely by un-forked hairs with spines are considered as poor pollen adhesive body. They forage on different species of flowers during their same foraging flight and are therefore considered as poor floral constancy pollinators and sometime predate other pollinators also. This attribute disturb the foraging behaviour of other pollinators.

Coleoptera- Beetle pollination (Cantharophily)

Cantharophily first time reported in a Brazilian asclepiad, involving the melyridae, *Astylus variegatus* and the nectariferous flowers of *Oxypetalum banksii.* Coleopterans are considered to be the most primitive insects as well as pollinators with biting and chewing type mouthparts and apparently primitive instincts, which generally preclude their effective participation in the pollination process. Usually beetles preferred large, greenish or off-white in color and heavily scented flower. Most beetle-pollinated flowers are flattened or dish shaped, with pollen easily accessible, although they may include traps to keep the beetle longer and well protected ovaries from the biting mouthparts of the pollinators. Beetles are unable to land precisely on flowers due to their morphology and flying habits, hence most anthophilous species are restricted in their visits to the simplest flowers. In such flowers, the beetles imbibe both nectar, pollen and chew the floral parts, especially pollen. The morphological evolution of flower visiting coleopterans have adaptations; forward projection of the mouth parts by tilting up of the head and associated prolongation of the pro-thorax and the neck. Maxillary setae may also be elongated. These evolution allow the proboscis of beetles to reach deeper nectarines to easily imbibe nectar e.g. Cerambycidae.

Diptera- Fly Pollination (myophily)

Flies are considered as less important, ineffective and unreliable pollinators but nonetheless, their absolute numbers and presence of some flies throughout the year make them important pollinators in some temperate and many tropical flowering plants. There are huge numbers of flies recorded as flower visitors which is an attribute of pollinators and indeed, proof of their importance as pollinators. A wide variety of flies feed upon pollen; notoriously the flower flies (Syrphidae). Flies are important pollinator at high altitude and high latitude where other insect pollinator groups are lacking. The adult of dipterans have suctorial and lapping type mouth parts which are also considered to be primitive pollinators. Some different flies of the Dipterans have been observed with a number of loose pollen adhered on their body and deposited on the stigma which is key character of true pollinator. This may be simply accomplished within the same flower or on the same plant. Nectar is the important and rich food source for fuelling activities of flies. About two dozen families; house flies and their relatives are foragers (Larson *et al.,* 2001). Some of these species are important in crop pollination, such as for mango and for horticultural crops. (Free, 1993). Most of the flies belonging to the suborder Nematocera have short suctorial and lapping mouth parts that restrict them to feeding on exposed nectar from open flowers such as roses, euphorbs and carrots (Larson *et al.,* 2001). Cacoa (*Theobroma cacao*) is pollinated by several kinds of midges which belong to families Ceratopogonidae and Cecidomyiidae (Roubik, 1995).

Lepidoptera

Lepidopteran pollinators may be divided in to two groups; butterfly and moth

Butterfly pollination (Psychophily)

Butterfly pollination is known as psychophily. It is diurnal in nature and has good vision to perceive a wide spectrum of colors with excellent sense of smell. They have chemoreceptor sense of smell at the ends of their antennae. Since they do not digest pollen (with one exception), more nectar is preferred than pollen. They perch high on their long thin legs and do not pick much pollen on their body due to the lack of specialized morphology. They have long and slender proboscis which can easily reach till the nectarine of those flowers with long cylindrical basal part. From these characteristics, one can easily recognize the characteristics of typical butterfly-pollinated flowers.

Moth Pollination (Phalaenophily)

Moth pollination is known as phalaenophily. It is nocturnal in nature. They have strong sense of olfaction and are active fliers which help to hover in front of flower without landing on the flower. Usually, they tend to be attracted towards white colour, night opening, and large tubular corollas where nectarines are deeply hidden. Moths are attracted to strong sweet scent producing flowers during the night *viz.* morning glory, tobacco, gardenia, yucca, dragon fruit, etc.

They are highly specialized flower visitors equipped with a long, thin and very flexible proboscis; during feeding, it stretches out to take nectar but keep it coiled up at other times. The tongue is usually extended just as the moth reaches its nectarine. The long and narrow tube shape corolla features are not suitable for short-tongued visitors. In the phalaenophily, Hawk-moth (sphinx moth) considers best pollinator for long and narrow shaped corolla based flower.

Thrips

Thrips are also considered pollinators because they carry loose pollen grains on their body and transfer pollen from flower to flower of the same plant and sometimes other plants. The loose pollens are carried by both adults and the larvae, where adults are more efficient than larvae due to their stature and greater surface area such as wing fringes, abdominal setae as well as the antenna. The active involvement of Thrips in pollination is well documented on Cocoa,

Capsicum, Onion, etc. In some plant species Thrips alone contributes 93 per cent pollination but usually they consider as poor or secondary pollinator.

Ants (Myrmecophily)

Usually ants are considered as poor pollinator. Although they are commonly anthophilous, their morphological characteristics are a constraint and prevent them from effectively participating in the pollination process. Pollen does not adhere readily on their hard and generally smooth bodies. The small body size also constantly deprives contact with the anthers and stigmas during foraging. The worker ants cannot fly and do not facilitate cross-pollination. Ants rob nectar from nectarines without effectual pollination so they consider as nectar robber. They usually visit inconspicuous, low growing flowers and those close to the stem. Until recently, no beneficial role in pollination was attributed to ants: they were just considered to be robbers of nectar and pollen. Apart from that, in Australia, ants as pollinators have been reported in *Microtis parviflora*. Jones (1975) reported that the worker ants of three different genera (*Iridomyrmex* sp.) (Dolichoderinae), *Meranops* sp. (Myrmeciinae); and *Rhytidoponera tasmaniensis* (Ponerinae) forage for nectar on (*Microtis parviflora*) orchid species. All three ant species effectively pollinate the *Microtis* blossoms, where *Iridomyrmex* sp. is the most frequent visitor.

Pollination by Other Insects

Some insects belonging to various orders are also reported as flower visitor, nectar and pollen feeder but while their attributes seems to be that of pollinator, they were not considered as a key pollinators. The pollinating effectiveness of many of these invertebrate groups is probably negligible. Apart from the above described invertebrates, flower-visiting taxa have been recorded amongst the following invertebrate orders in various parts of the world (Muller, 1883; Proctor and Yeo, 1973): Collembola as a pollen feeder, Mecoptera as a pollen and nectar feeder, Neuroptera as a pollen and nectar feeder, Odonata as a pollen and nectar feeder, Plecoptera as a pollen and nectar feeder, Dermaptera as a pollen and nectar feeder, Dictyoptera as a nectar feeder, Hemiptera as a flower visitor, Psocoptera as a flower visitor, and Trichoptera; snail and slug are flower visitors and their pollination known as Malacophily. Orthoptera: *Zaprochilus* and a closely related genus of "Zaprochilinae" feed on the pollen and nectar of flowers.

References

Abrol, D .P., 2012. Pollination biology- biodiversity conservation and agricultural productivity. Springer Dordrecht Heidelberg London New York, 787 pp.

Aizen, M., Garibaldi, L., Cunningham, S. and Klein, A. 2008. Long-term global trends in crop yield and production reveal no current pollination shortage but increasing pollinator dependency. Current Biology, 18:1572-1575.

Buchmann, S. L. and Ascher J. S. 2005. The plight of pollinating bees. Bee world, 86:71-74.

Burd, M. 1994. Bateman's principle and plant reproduction: the role of pollen limitation in fruit and seed set. Botanical Review, 60: 83-139.

Chacoff, N. and Aizen, M. 2006. Edge effects on flower-visiting insects in grapefruit plantations bordering premontane subtropical forest. Journal of Applied Ecology, 43: 18-27.

Daily, G. C., Alexander, S., Ehrlich, P. R., Goulder, L., Lubchenco, J., Matson, P. A., Mooney, H. A., Postel, S., Schneider, S. H., Tilman, D. and Woodwell, G. M. 1997. Ecosystem services: benefits supplied to human societies by natural ecosystems. Issues in Ecology: 1-16.

FAO. 2007. Pollinators: Neglected biodiversity of importance to food and agriculture CGRFA-11/07/Inf.15. Food and Agriculture Organization of United Nations, Rome.

FAO. 1993. Harvesting nature's diversity. FAO, Rome.

Free, J. B. 1993. Insect pollination of crops, 2nd edn. Academic, London.

Gallai, N., Vaissie`re, B. E., Potts, S. G. and Salles, J. 2009. Assessing the monetary value of global crop pollination services. Oxford University Press.

Jones, D. L. 1975. The pollination of Microtis parviflora R. Br. Annals of Botany, 39:585-589.

Klein, A-M., Vaissi-Are, B. E., Cane, J. H., Steffan-Dewenter, I., Cunningham, S. A, Kremen, C. and Tscharntke, T. 2007. Importance of pollinators in changing landscapes for world crops. Proceedings of the Royal Society B: Biological Sciences 274: 303-313.

Larson, B. M. H., Kevan, P. G. and Inouye, D. W. 2001. Flies and flowers: taxonomic diversity of anthophiles and pollinators. Canadian Entomology, 133:439-465.

Linsley, E. G. 1958. The ecology of solitary bees. Hilgardia, 27:543-599.

Michener, C. D. 2000. The bees of the world. The John Hopkins University Press, Baltimore/ London, 913 pp.

Michener, C. D. 2007. The bees of the world, 2nd Edn. The Johns Hopkins University Press, Baltimore.

Nabhan, G. P. and Buchmann, S. 1997. Services provided by pollinators. In G. C. Daily (Ed.), Nature's services: Societal dependence on natural ecosystems (133-150). Washington, DC: Island Press.

Pritchard, K. 2005. The unseen costs of agricultural expansion across a rainforest landscape: Depauperate pollinator communities and reduced yield in isolated crops. Unpublished master's thesis, James Cook University, Queensland, Australia.

Roubik, D. W. 1989. Ecology and natural history of tropical bees. Cambridge University Press, New York.

Roubik, D. W. 1995. Pollination of cultivated plants in the tropics. Food and Agricultural Organization of the United Nations, Rome, Bulletin 118.

2

Principles of Pollination

Introduction

Pollination is a process in which pollen grains are transferred from male gametes to the female gamete, thereby enabling fertilization and reproduction through growth of the pollen tube and eventual release of sperm. Fertilization is the union of the egg and sperm. In plants, this may happen after pollination; however, pollination doesn't always result in fertilization. Fertilization occurs when the pollen grains germinate on the stigma and grow in the style. Reproduction is the process to produce offspring by fusion of male and female gametes, which simply means that the sperms of the pollen unite with the ovules in the ovary of the flower and subsequently produce seed. Flowers may facilitate out-crossing (fusion of sperm and eggs from different individuals in a population) or allow selfing (fusion of sperm and egg from the same flower). Some flowers produce diasporas without fertilization (parthenocarpy).

The requirement of pollinator depends upon the type of pollination required by plant species and varieties. Whether it is self-fertile and partially self-pollinating, or self-fertile and not self-pollinating, or self-infertile. The efficiency of pollinators may vary according to pollinator species, foraging rate, time spent, loose pollen carried and pollen deposition on stigma. Dependency of crops on pollinators also varied, some crops give no increased yield or only a moderately increased yield when pollinated by insects, some give a greatly increased yield, and some little or no fruit or seed produce without pollinators' pollination. Unfortunately, the pollination requirements of some crop species and cultivars are unknown, and key pollinators of many crops are unknown. Research is needed to find whether the particular crop benefits from self-pollination or cross-pollination, which pollinator species are key pollinator of the flowers, and whether fruit or seed set is usually adequate or suffering from deficit of pollination. A comprehensive study is required to list of key pollinators of particular crops in particular region and pollinators ranking on the base of their pollination efficiency.

Pollination: Transfer of pollen grains from anther to stigma of the same species is known as pollination.

Types of Pollination

In angiosperm, two types of pollination may occur: self-pollination and cross pollination.

Self-Pollination

Self-pollination refers to the transfer of pollen from anther to stigma of the same individual plant to accomplish fertilization. There are two types of self-pollination

Autogamy: It refers to transfer of pollen from the anther to the stigma of the same flower

Geitonogamy: It refers to transfer of pollen from the anther of one flower to the stigma of another flower of the same plant.

Cross-Pollination (allogamy)

It refers to the transfer of pollen from the anther of one plant to the stigma of the flower of another plant in order to accomplish fertilization.

Biotic and abiotic agents play crucial role in cross-pollination, in which insects and wind are key agents. Out of biotic agent, insects are key agent and out of insects honeybees are the most important and efficient pollinator.

Cleistogamy

Cleistogamy or closed marriage refers to pollination and fertilization (autogamy) occurs before the anthesis of flower and always self-pollination occurs. Example- peanuts, some peas, some beans etc. Some cleistogamous flowers never open.

Chasmogamy

Chasmogamy or open marriage refers to fertilization occurs after anthesis of flowers. However, pollination may occur before or after the anthesis of flowers. When chasmogamous flowers have reached maturity, they unfurl either their stamens or style or both for pollination. Most of the chasmogamous flowers are cross-pollinated by biotic or abiotic agents although some plant species possess self-fertilization traits also.

Type of Flowers' sexuality

Bisexual flower

Bisexual flower is a perfect flower that has both male (androecium) and female (gynoecium) reproductive structures. It also known as hermaphroditic flowers.

Unisexual Flower

Unisexual flower have either functionally male or functionally female reproductive structure. In angiosperms this condition is also called imperfect, or incomplete flower.

Angiosperm

The flowering plants are known as angiosperms and it is also known as angiospermae.

Gymnosperms

The term "gymnosperm" comes from the two Greek word, *gymnos* means naked and *sperma* means seed, meaning naked seeds. Gymnosperm seeds develop either on the surface of scales or leaves, often modified to form cones, or at the end of short stalks as in Ginkgo. The gymnosperms are a group of seed-producing plants that includes Conifers, Cycads, Ginkgo, and Gnetales.

Type of Plants' Sexuality

Monoecious Plants

It is a Greek word which means one household. Monoecious plant refers to those plants in which male and female flowers are separate but bloom on the same plant e.g. maize, cucumbers, etc.

Dioecious Plants

It is a derived from a Greek word that means two household. Dioecious plants refer to those plants in which male and female flowers develop on different plants *viz.* kiwi, willow, poplars, etc.

Androecious plant

These plants produce only male flowers and pollen but no seeds. The male plants occur in a dioecious population.

Gynoecious Plant

These plants produce only female flowers without pollen and seeds without pollination. The female plants occur in a dioecious population. In some plant populations, all individuals are gynoecious with non sexual reproduction used to produce the next generation. Examples; few variety of cucumber and cactus, etc.

Hermaphrodite Flower

It is a synonym of a Greek word monoclinous that means one bed. Hermaphrodite refers to a flower which has bisexual reproductive flower. Simply speaking, Hermaphrodite flowers bear that type of flower which has both staminate (male, pollen-producing) and carpellate (female, ovule-producing) parts. A closer analogy to hermaphroditism in botany is the presence of separate male and female parts on the same flower and such flower are called monoecious.

Dichogamous plant These plants having both sexes androecium and gynoecium may mature at different times; viz. protandrous or protogynous. Thereby Dichogamous offers self-pollination (Stace, 1995).

Protoandrous Plant

It refers to the male part (anthers) matures before the female part (stigma) on the plants.

Protogynous plant

It refers to the female part (stigma) matures before the male part (anthers) on the plants.

Self Compatible

Those pollens that are capable and accomplish fertilization thereby, form seed and fruit set with its own pollen are known as self compatible flower of the plants. Ex. Pigeonpea, tomato, chili, etc.

Self in-compatible

It refers to the failure of pollen from a flower to fertilize the same flower or other flower of the same plant. Ex. Aonla, apple, etc.

Pin flower: It refers to those flowers have long styles (stigma) and short stamen (anther).

Thrum flower: It refers to those flowers have short styles (stigma) and long stamens (anther).

Anthesis

Anthesis is the period during which a flower is fully open and functional.

Anthers Dehiscence

Anthers dehiscence refers to the splitting of the anther at maturity and release of the pollen grains.

Flower Structure

Usually, flower parts are divided in two groups: the vegetative part and the reproductive part. The vegetative parts of flower consist of perianth, calyx and corolla. These are vegetative parts but still important because, usually, calyx protects the flower when it is still a bud and corolla helps to attract pollinator and land on the flower. The reproductive parts consist of androecium and gynoecium. The perianth is the vegetative part of the flower and this structure forms an envelope or tube-like tissue surrounding the sexual organs, consisting of the calyx and the corolla. Calyx is the outermost whorl consisting of units called sepals; these are typically green and enclose the rest of the flower in the bud stage. However, they can be absent or prominent and be petal-like in some species. Corolla is the next whorl toward the apex, composed of units called petals, which are typically thin, soft and colored to attract animals that help the process of pollination. Androecium is a Greek word and comprise of two words (andros oikia: man's house) which is the next whorl consisting of units called stamens. Stamens consist of two parts: a stalk called a filament, topped by an anther where pollen is produced by meiosis and eventually dispersed. Gynoecium is a Greek word and comprise of two words (gynaikos oikia: woman's house) which is the innermost whorl of a flower, consisting of one or more units called carpels. The carpel or multiple fused carpels form a hollow structure called an ovary, which produces ovules internally. Ovules are megasporangia and they produce megaspores by meiosis which develops into female gametophytes. These give rise to egg cells. The gynoecium of a flower is also described using an alternative terminology wherein the structure in the innermost whorl (consisting of an ovary, style and stigma) is called a pistil. A pistil may consist of a single carpel or a number of carpels fused together. The sticky tip of the pistil is the stigma which exerts itself as the receptor of pollen. The supportive stalk, the style, becomes the pathway for pollen tubes to grow from pollen grains adhering to the stigma. The relationship to the gynoecium on the receptacle is described as hypogynous (beneath a superior ovary), perigynous (surrounding a superior ovary), or epigynous (above inferior ovary).

Many crops – apples, citrus, peaches, pears, plum, sunflowers, cabbage, cauliflower, and mustard produce hermaphrodite (bisexual) flowers that have both male and female sex organs. However, there are crops, various cucurbits produce monosexual (either male or female) flowers on the same or different branches of the same plant. The kiwi fruit produce male and female flowers on different plants. Some flowers may be fully pollinated but not necessarily fertilized because they have received incompatible pollen. Therefore, the pollen does not germinate or grow on the stigma or reach and fertilise the ovules. In such a case, unless the species is parthenocarpic, no fruit will form. Not all plants require pollination and fertilisation, as some are able to produce fruit parthenocarpically; that is, the fruit will develop without fertilisation of the flower and production of seed.

Ovary

Ovary is a part of the female reproductive organ of the flower or gynoecium. Specifically, it is the part of the pistil which holds the ovule(s) and is located above or below or at the point of connection with the base of the petals and sepals.

Nectaries and Nectar Secretion

Flowers frequently have one or more *nectarines.* The nectary is most often located within the flower, usually at the base of the sexual column inside the circle of petals. However, in cotton the nectariferous ring lies just outside the base of the petals on the inner base of the calyx. Nectaries are also found outside the flower, on the stem or leaves viz. rubber plant. Nectar secretion within the flower usually starts about the time the flower opens and ceases soon after fertilization. This secretion of nectar on the leaves is not influenced by flowering time and may continue for several weeks. The amount of nectar secreted varies from infinitesimal in numerous species to more than an ounce in the orchid *Coryanthes* spp.(Kerner 1878). Numerous scientists have determined the amount of nectar produced in the flowers of various crops (Todd and Mcgregor 1960).

Stigma Receptivity

When the stigma is receptive than it appears shiny and glow due to the coating of a colorless, relatively tasteless stigmatic fluid. If viable, compatible pollen comes in contact with this moist stigma, adheres to it, germinates, and create pollen tube and the sperm nuclei reach up-to the style into the ovary and, finally, into one of the ovules to accomplish fertilization.

Pollen Viability and Longevity

Pollen is an important vector of gene flow in plants, particularly for outcrossing species like tall Fescue. Viable pollens are able to germinate under favorable conditions. Even a glimpse of the vast literature on pollination ecology and pollen biology shows a wide variety of terms and definitions regarding pollen functional ability. Viability has been defined as "having the capacity to live, grow, germinate or develop" (Lincoln *et al.* 1982). But viable pollen grains may not actually germinate (in vitro or in vivo) if the conditions are not right. Pollen longevity will be used here to denote the period in which pollen remains able to germinate on the appropriate (receptive and compatible) stigma.

Pollenizer

The terms "pollinator" and "pollenizer" are often confused. Pollenizer is the plant that provides the pollen. Pollinator is the agent that transfers the pollen *viz.* bees, bats, moths, birds etc.

Out-crossing

Out-crossing or allogamy refers to cross-fertilization, in which offspring are produced by the fusion of the gametes belonging to two different plants of same species. This is the most common mode of reproduction among higher plants. About 55% of higher plants species reproduce by out-crossing and additional 7% are reproduced by partial allogamy, cross-fertilization (allogamy) and partially self-fertilizing (autogamy) process. The various aspects of floral character and morphology promote out-crossing. The primary mechanism used by flowering plants to ensure out-crossing is known as self-incompatibility. The bisexual flowers, the anthers and carpels (female reproductive organ of a flower) may mature at different times, plants being protandrous (anthers mature before carpels) or protogynous (carpels mature before anthers). Monoecious species, with unisexual flowers on the same plant, may produce male and female flowers at different times. Dioecy is the condition of having unisexual flowers on different plants which necessarily results in out-crossing, and might thus be thought to have evolved for this purpose.

Pollinators (Pollinating agent)

The process of pollination requires a carrier to carry or transfer the pollen grains from the anther to stigma. The carrier may be biotic or abiotic. Pollinators refers to those biotic agents that transfer pollen from the anthers to the female stigma of flower of the same species in order to accomplish fertilization. Faegri and van der Pijl (1979), in their classic book on pollination ecology, defined two major types of pollen dispersal: Biotic pollination in which the pollen dispersal agent is an animal (i.e., either an invertebrate or a vertebrate); and abiotic pollination where pollen is dispersed by an inanimate physical agent, such as wind or water. The various flower traits that attract different pollinators are known as pollination syndromes. The methods of pollination, with common pollinators or plants, are:

Biotic Pollinating Agent and Their Pollination

Invertebrates

Coleoptera: Beetles: Cantharophily

Diptera: Flies: Myophily

Hymenopterans: Wasps: Sphecophily

Bees: Melittophily

Ants: Myrmecophily

Lepidopterans: Butterflies: Psychophily

Moths: Phalaenophily

Vertebrates

Birds: Ornithophily

Mammals: Therophily

Bats: Chiropterophily

Rodents: Sminthophily

Abiotic Pollinating agent and their pollination

Self-pollination: Autogamy + Cleistogamy

Wind-pollination: Anemophily

Water-pollination: Hydrophily

Gravity pollination: Geophilly

Abiotic pollinating agents

Anemophily/ Wind pollination

Anemophily or wind pollination is a kind of abiotic pollination whereby pollen is carried by wind velocity. Most of the gymnosperms are anemophilous *viz.*

grasses, sedges, rushes, pines, firs, spruces, etc. Other common anemophilous plants are oaks, sweet chestnuts, etc. Anemophily is the dominant type of abiotic pollination and these plants must release huge pollens as possible. Anemophilous pollen grains are light weight and non-sticky, so that they can be transported by air currents. Pollen from anemophilous plants tends to be smaller and lighter in weight than pollen of entomophilous, with very low nutritional value to insects. However, insects sometimes gather pollen from flowers *viz.* maize and paddy during floral dearth. The anemophilous plants have morphological adaptations that increase pollen dispersal and capture. The male and female reproductive organs are generally found in separate flowers, the male flowers having a number of long filaments terminating in exposed stamens, and the female flowers having long, feather-like stigmas. This separation not only prevents self-fertilization and increases outcrossing but prevents stigmas from being clogged by self pollen. Anthers generally do not open unless the weather is favorable, i.e. warm and dry, because pollen is rapidly washed out of the air in rain. The female portion of anemophilous flowers has evolved to capture and utilize wind carried pollen. One common arrangement of anemophilous flowers is to be located higher than male flowers on the plant so that pollen will not just fall down onto stigmas of the same plant. At the same time, because of the high inefficiency of wind pollination, anemophilous species produce huge numbers of pollen grains.

Hydrophily/ Water Pollination

Hydrophily is a uncommon pollination process, whereby pollen is distributed by the flow of waters, particularly in rivers and streams. It is highly specialized mechanism that occurs rarely among aquatic angiosperms, which mainly retain reproductive system reminiscent of their terrestrial ancestor. In some flowers, the pollen is released into the water and floats on the water surface. The female flowers emerge on the surface, receive pollen, and are then withdrawn back under the water, viz. Hydrilla, Callitriche, Ruppia, Zostera, Elodea, etc. In Vallisneria, the male flowers become detached and float on the surface of the water; the anthers are thus brought in contact with the stigmas of the female flowers.

Geophily/ Gravity Pollination

Geophily or gravitational pollination is a kind of biotic pollination whereby pollens are transferred by the gravitational force of the earth. It is found in self pollinated plant species. In this case, pollen falls down because of gravitational force on to the receptive stigma of other flowers. However, gravity is highly unreliable and a rare and insignificant pollinating agent.

Biotic Pollinating Agents

Pollination by biotic agents is known as zoophily. Biotic pollinating agent (pollinators) refers to those living organisms which involve in transfer of pollen grains from anther to stigma viz. bees, wasps, butterflies, moths, flies, bats, birds, etc. The biotic agents visit on flowers either for nectar or pollen or both, and they incidentally transfer pollen grains from anther to stigma of one flower to another flower of the same or another plant. Plant and pollinators relationship is a special kind of symbiotic relation; mutualism, which is co-evolving in nature. Most plant species have adopted different kinds of relationships with biotic pollen vectors (Herrera 1996). One option is to be a generalist and try to attract a wide variety of different pollinators. The other is to specialize (and often coevolve) with a single type of pollinator. Plants which are specialists, can adapt to very specific and highly efficient pollination mechanisms but are restricted to co-occurring with their pollinators. It is these mutualists that draw the most attention in pollination biology with their intricate and peculiar mechanisms of pollination. Whether adapted as generalists or specialists, zoophily plants have several characteristics. The pollen is sometimes larger than in anemophilous species, is often sticky and/or highly ornamented with spines and bumps, and sometimes adheres in clumps of several grains. The number of pollen grains, sometimes expressed as a ratio with the number of ovules, is much lower in biotic pollination than in anemophilous species. Stamens are located so as to contact the pollinators, rather than to be exposed to wind. However, much of the floral structure is related to attracting specific types of animals to act as pollinators.

Pollination by birds is known as ornithophily. It is very common in South America and Australia; whereas in Asia, it occurs on few plants in the Hindu Kush- Himalayan region. Birds visit only those few plant species, those that produce plenty of nectar, *viz.* avocado, visited by humming birds. Mammals are, however, the pollinating agents of only a few plants (Ayensu 1974). Some mammals bat, Honey possum and Namaqua rock mouse visit on flowers of a particular plant species for nectar, and thereby accomplish pollination.

Insect pollination is known as entomophily and plants that require insect pollination to bear fruit and seed are called entomophilous. It is dominant in ecosystem and found in many agricultural crops, horticultural crops, forage crops, ornamental plants, and other wild plants which are all effectively pollinated by insects that visit flowers either for nectar or pollen or both. Insect pollinators pollinate about 80% of flowering plants (FA0, 2007). Different kinds of insects such as bees, flies, beetles, butterflies, moths, beetles, weevils and wasps are important pollinators of many plants. Bees are one of the most effective and reliable pollinators.

Pollination Syndromes

Pollination syndrome as a suite of floral traits, including rewards is associated with the attraction and utilization of a specific group of animals as pollinators (Fenster *et al.* 2004). Pollination syndromes are suites of traits and trait combinations that are hypothesized to increase the attraction and pollen transfer of specific types of pollinators (Reynolds *et al.*, 2009). In simple term, pollination syndromes are suites of flower traits that have evolved in response to natural selection imposed by different pollen vectors, which can be abiotic (wind, water and gravitational) or biotic (insect, birds, bats, etc.). These traits include flower shape, size, colour, odour, reward type and amount, nectar composition, volume of nectar and pollen, timing of flowering, etc. For example, tubular red flowers with copious nectar often attract birds; foul smelling flowers attract carrion flies or beetles, etc. Floral evolution has often been associated with differences in pollination syndromes. The floral traits are expected to correlate with one another across independent evolutionary events.

Birds, bats and insects play a vital role in pollination of most fruits and vegetables. Fenster *et al.* 2009 observed four groups of visitors: 1) hummingbirds, 2) bees and wasps, 3) butterflies and moths and 4) flies. They found that the volume of floral nectar and size of the floral display was significantly correlated with transitions to hummingbird pollination, while length of corolla tube and flower color did not significantly correlate with transitions to any of the four major pollinator groups. Only two of four traits could be directly associated with transitions to specific pollination systems and concluded that the patterns do not fit those predicted by classical pollination syndromes.

Floral Rewards

Floral rewards are considered as those components of flower or inflorescence that is used by visitor, and due to their vital use ensure repeated visitation that will lead to pollination. The flower of plants attract pollinators and offer rewards such as as nectar, pollen etc. Nectar is a rich source of sugar that provides energy while pollen is a rich source of protein for development. The plant must offer reward in the form of something the pollinators need, either for their own survival or for that of their offspring and in the symbiotic process, the plant itself achive pollination. Reciprocally, pollinators transfer pollen grain from anther to stigma. This process was, for the first time, realized by Christian Sprengel (1750–1816) where the bees and other flower visitors do not provide their pollen-carrying services free of charge. Many other inducements are exerted to attract pollinators to specialize on a particular species (or genus) of plant existence. Flowers attract the pollinators by being conspicuous in color, scent, size, or shape, making it easier for visitors to pick them out from their surroundings, where the true floral rewards are nectar and pollen.

Pollen Reward

Pollen may have been the original reward by which most of the flowers began attracting insects in the Cretaceous period. Pollen is the male gametophyte and flowers always produce numerous pollens but out of these, only few pollen grains are used for fertilization. Pollen is a highly nutritious food containing protein, carbohydrate, fat, minerals, antioxidants, amino acids and vitamins. Plants that offer pollen as a reward often produce it in huge quantities. In order to minimize the loss of viable pollen to pollinators, some plants produce, in the same flower, two types of anthers, normal ones, which produce healthy pollen, and others that yield sterile or at least less viable but still very nutritious and probably tasty pollen.

Pollen itself has been shown to "jump" from the stamens onto the insect, then from the insect onto a receptive stigma. It is known that electrostatic charges can build up on the body of a foraging honeybee and that these charges may reach a magnitude of several hundred volts. The bee is therefore flying at the center of its own highly charged electrostatic field. Most floral surfaces are well insulated, but the pistil is an exception and indeed, a path of very low resistance leads from the stigma to the ground, almost like the earthing track of a lightning conductor. The result is that the bee's electrostatic field is attracted to the stigma, and with it the pollen.

Nectar Reward

Nectar is also an important floral reward secreted by nectarine of the flower. Many of the plants offer nectar but not all flowering plant species. Pollinators must determine whether a floral scent is always associated with nectar. The ability of insect pollinators to learn floral scents shapes the nature of floral scent in a way that keeps plants honest about the presence of nectar. Nectar is an ideal substance to offer as a reward (Baker and Baker 1990). It is easily produced by plants; it can be produced quickly as the demand requires, and, being a sugar solution, it is readily digested and quickly assimilated as a source of energy. However, nectar has been modified in some species to reward specific pollinators. One critical factor is the amount of nectar, as flowers "want" to reward pollinators without satiating them. If too much nectar is provided, the pollinator has no need to forage further, and plant may face pollination deficits. On the other hand, flowers of different species, and even within species, are competing with one another for pollinators, and those that provide higher volume or high concentration of nectar to attract abundant pollinators. The nectar can vary in concentration and content. Ornithophilus (bird pollinated) flowers usually have a fairly dilute nectar, with a sugar concentration between 20 to 30 per cent. On the other hand, horse-chestnut flowers may have a sugar concentration as high as 70%, with the results that bee visitors have to dilute it with their own saliva before they can suck it up (Baker et al. 1998). The observation that certain nectars and consequently some honey, can be poisonous has received a great deal of attention. The *Rhododendron ponticum* and other species are also known to produce poisonous nectar or pollen.

Aesthetic Reward

The primary attractants of flowers include color, size, shape, and scent. Plants use pigments contained in vacuoles to generate the huge variety of flower colors. Virtually, the purpose of floral color exerted is to attract pollinators. Plants combine these pigments, along with special surface textures, to create the many additional colors, as well as shiny and velvety appearances. Most of the flowers produce scents to attract pollinators. Many classes of pollinators have keen senses of smell. Their preferences vary from pollinator to pollinator, as does their color vision, and the scents given off by flowers reflect these differences. In many orchids, the scent is so specific and therefore, they have specific pollinators. Flowers of different species are distinguished more by their smells than by their colors, simply because there are many, many odor categories and relatively few colors. For this reason, odor may be even more important than color in establishing a phenomenon known as flower constancy. This flower constancy shows loyalty of pollinator towards the flower of plant species.

The flowers of plants are multi-sensory advertisements which lure pollinators to come into contact with a plant's reproductive organs. In turn, pollinators rely both on the visual and olfactory cues provided by these advertisements to locate and identify flowers with food resources such as nectar, pollen and oils. Honeybees are able to sense high concentration odours faster than low concentration odours (Wright and Smith 2004) and can learn to identify individual compounds in complex floral scents consistently associated with reinforcement (Smith and Cobey 1994; Wright and Smith 2004). The floral color is important for recognition and attraction of the flora from long-distance while floral morphology exert an attractant as well as a determinant of the effectiveness of pollination by pollinators and floral rewards are the ultimate reason for visits by pollinators.

The evolution of flowering plants has undoubtedly been influenced by a pollinator's ability to learn and associate floral signals with food. The pollinator-mediated selection for floral scent production should favour signals which are distinctive and exhibit low variation within species because these signals are learned faster. Social bees quickly learn to associate scent with the presence of nectar, and their ability to do this is generally faster and more reliable than their ability to learn from visual cues and as a result, pollinators rely on floral scent as a means of distinguishing honestly signalling flowers from deceptive ones (Wright and Schiestl 2009). Hummingbirds has shown no color preferences among red, yellow, lavender- pink, light pink, medium pink but foraged according to floral shape and bumblebees showed no floral shape preferences but foraged according to floral color, they avoided red flowers but visited yellow, lavender-pink (Sutherland and Vickery).

Other Reward

Many other types of rewards are offered in the great diversity of types of flowers. Some flowers produce fatty oils (glycerides) to offer visiting bees instead of nectar. The bees that pollinate these species have specialized hairs to collect, transport, and temporarily store the somewhat sticky oil. In general, it is only the female bees that collect the oil, and it is they who mix the collected oil with pollen to produce a nutritious and protein-rich "bee-bread" for their larvae. Other species produce resins and waxes, which can be valuable as nest-building materials, especially in wet-tropical areas while some flowers may provide heat as a reward (Stanton and Galen 1989).

References

Ayensu ES (1974) Plant and bat interactions in West Africa. Ann Missouri Bot Gard 61:702–727

Baker HG, Baker I (1990) The predictive value of nectar chemistry to the recognition of pollinator types. Isr J Bot 39:157–166

Baker HG, Baker I, Hodges SA (1998) Sugar composition of nectars and fruits consumed by birds and bats in the tropics and subtropics. Biotropica 30(4):559–586

FA0. 2007. Pollinators: Neglected biodiversity of importance to food and agriculture CGRFA-11/07/Inf.15. Food and Agriculture Organization of United Nations, Rome.

Faegri K, van der Pijl L (1979) The principles of pollination ecology, 3rd edn. Pergamon, Oxford

Fenster, C.B., Armbruster, W. S. Wilson, P. Dudash, M. R. and Thomson, J.D. 2004. Pollination syndromes and floral Specialization. Annu. Rev. Ecol. Evol. Syst. 35:375-403.

Fenster, C.B., Martén-Rodriguez, S., Schemske, D. W. 2009. Pollination Syndromes and the Evolution of Floral Diversity in Iochroma (Solanaceae). Evolution doi: 10.1111/j.1558-5646.2009.00730.x

Herrera CM (1996) Floral traits and plant adaptation to insect pollinators: a devil's advocate approach. In: Lloyd DG, Barrett SCH (eds) Floral biology. Studies of floral evolution in animal pollinated plants. Chapman & Hall, New York, pp 65–87

Kerner A (1878) Flowers and their unbidden guests. The translation revised, and edited by w.Ogle. Kegan, London. 164 pp

Lincoln R. J., Boxsh*all G. A.*, Clark P. F. (1982) A dictionary of ecology, evolution and systematics. Cambridge University Press, New York.

Reynolds, R.J., Westbrook, M.J., Rohde, A.S., Cridland, J.M. Fenster, C.B. and Dudash, M.R. 2009. Pollinator specialization and pollination syndromes of three related North American Silene. Ecology, 90(8):2077-2087.

Smith, B.H. & Cobey, S. (1994) The olfactory memory of the honeybee Apis m*ellifer*a blocking between odorants in binarymixtures. Journal of Experimental Biology, 195, 91-108.

Stace, H.M. (1995). "Protogyny, Self-Incompatibility and Pollination in Anthocercis gracilis (Solanaceae)". Australian Journal of Botany 43 (5): 451-459. doi:10.1071/BT9950451

Sutherland, S.D. and Vkkery, R.K. On the relative importance of floral color, shape, and nectar rewards in attracting pollinators to Mimulus. Great Basin Naturalist 53(2):107-117.

Todd FE, Mcgregor SE (1960) The use of honey bees in the production of cr*ops. Ann Rev E*nt 5:265–278

Wright, G.A. & Smith, B.H. (2004) Variation in complex olfactory stimuli and its influence on odour recognition. Proceedings of the Royal Society of London B-Biological Sciences, 271, 147-152.

Wright, G.A. and Sc*hiestl,* F.P. 2009. The evolution of floral scent: the influence of olfactory learning by insect pollinators on the hon*est signaling of flor*al rewards. Fun*ctional Ecology, 23, 841–851.* **d**oi: 10.1111/j.1365-2435.2009.01627.x

3

Population Depletion and Extinction of Insect Pollinators

Pollination is a vital and key component of ecosystem services which is provided by pollinators. FAO, 2007 reported that 80 per cent of plant species rely for pollination on pollinators, which is dominated by insects. Among the insect pollinators, solitary and social bees dominate both managed and natural ecosystems. No other groups of insects are more beneficial to humans than bees. Many plants have evolved intricate relationships with many insect pollinators and without their presence, majority of plant species would not reproduce and maintain their genetic diversity (Daily *et al.*, 1997). In natural ecosystem, insect pollination plays a pivotal role to maintain and conserve the plant biodiversity. On a global scale, the total annual value of insect pollination services has been estimated at USD 217 billion (Sciencedaily, 2008). United Nations reported that the economics of ecosystems and biodiversity by insect pollination was valued at £134 billion.

Both wild and domesticated insect pollinators are declining and in parallel those plants that are relying for pollination upon them are also declining gradually. Insect pollinators and other pollinator species are both declining at an alarming rate which is a threat to the existing plant pollination and the life of their progeny. This depletion trend could loss dozens of commercially important crops. The declining pollinator population and diversity is a serious concern to agricultural production, retention and conservation of biodiversity in many parts of the world. Depletion of pollinator population create scarcity of pollination services which have negative ecological and economic impacts that could significantly affect the plant diversity, wider ecosystem stability, crop productivity, food security and human welfare. The worldwide pollinators are declining due to habitat loss, habitat fragmentation, agrochemicals, monoculture, pathogens, enemies, exotic species, climate change and their interactions.

Although the *Apis mellifera* has been widely accepted as one of the most important pollinators, however, population of honey bees are declining rapidly, causing global concern for pollination services (Biesmeijer *et al.,* 2006). The extent of our reliance on single species for such an important service is risky. In the United States, managed stocks of the honey bee have declined by 50 per cent due to CCD (Colony collapse disorder) over the past few decades (National Research Council, 2007). Loss of bees could seriously threaten productivity and biodiversity (FA0, 2007). It is important to ensure a scientifically reliable source of pollinators with comprehensive strategy for management of crop pollination is needed.

Conservation and augmentation support of domestic and wild pollinators effectively secure pollinators for sustainable agriculture. Global convention of bio-diversity has identified the importance of pollinators and established the International Pollinators Initiative (IPI) in 2000 for the conservation and sustainable use of pollinators which is facilitated and coordinated by FAO. This organization has established a Global Action on Pollination Services for sustainable agriculture through technical assistance and perusal of pollination prospect.

Insect Pollinator Erosion

The evolution of pollinators has taken millions of years while their population erosion and extinction has become very fast. There is the need to understand the relationship between plant pollinator and pollination in order to quantify the requirement of pollinators for different cultivated crops and wild plants to optimize the pollination. There is also the need to define the term 'pollinator crisis' in broad spectrum and came to knowledge of those crops which are suffering from pollination crisis. Our dependence on a single pollinator for such an important service is risky and global organizations are recognizing the need of pollinators' abundance and diversity, particularly native species. Under ideal pollination conditions, there is the demand of asymptote and diversified pollination services by ecosystem. The ecological consequence of contemporary agriculture can be viewed from various angles analyzing each component of agriculture deforestation for expanding agriculture, habitat destruction, modern agricultural practices, monoculture, intensive tillage, heavy irrigation, injudicious uses of agro-chemicals, climate change and other agronomic practices with their influence on the ecosystem and thereby pollinators.

Studies on the health of pollinator species reveal an overall lack of consensus on both pollinator crisis and potential factors that influence pollinator's health. Those studies support the pollinator crisis hypothesis, warn of future large-scale losses of agricultural productivity due to the decline of pollination services. In the last two decades, there have been a number of examples observed globally; failure of crops directly due to the pollinator scarcity, *viz.* failure of pumpkins, cherries, alfalfa, blueberries, cashews and brazil nuts (Allen- Wardell *et al.,* 1998). On the other hand, some experts do not support the pollinator crisis hypothesis, but all the scientists recognized the importance of pollination services as also to proliferate their population, diversity and support to continuous intensive research and monitoring worldwide. The global pollinator crisis becomes apparent, pollinator crisis is driven mainly by reported declines of honey bees in North America, bumblebees and butterflies in Europe, whereas native pollinator species show mixed responses to environmental change. Nearly one third of Europe's 435 butterfly species are reported to be in decline (The Telegraph, 2010). The presumption of plentiful honey bees for crop and ecosystem pollination has been severely challenged in the past several years by mysterious declines of honey bee colonies (Aizen and Harder, 2009; van-Engelsdorp *et al.,* 2007). The rapid loss of adult worker bees in colonies and the lack of apparent symptoms has led to the imprecise labeling of 'Colony Collapse Disorder' (CCD). This disorder is not universal and, in fact, global populations of honey bees are increasing (Aizen and Harder, 2009).

It may be due to inadequate pollinators that has reduced crop produce quality (e.g. fruit size, number of seed and nutraceutical chemical composition). Insects, particularly bees, are the efficient pollinators of most agricultural crops and wild plants pollination services depend upon both domesticated and wild pollinator populations. Growing concern about the fate of both domesticated and wild pollinators has resulted in the establishment of special initiatives by the Convention on Biological Diversity (International Pollinator Initiative) and several continental, national and regional programmes to tackle the issues of pollinator recession.

The link between pollinators and global food security, pollinators and biodiversity and declining of domestic pollinators and wild pollinators are an increasing concern. Undoubtedly, the health of honey bees and many Non-Apis; solitary and social bees are declining (Ghazoul, 2005; Steffan-Dewenter *et al.,* 2005). Honey bee population, specially forager bees are badly affected by the injudicious use of pesticides upon blooming stage of flowers on bee foraging plants and the chronic exposure to acaricides which is used to control the parasitic mite *Varroa destructor*. Furthermore, destruction and fragmentation of their habitats, monoculture and intensive tillage have significant negative effects on honey bees and other pollinators (Kremen *et al.,* 2007) while climate change also affects their population. The honey bees are attacked by several natural enemies *viz.* parasitic mites (*Varroa destructor*, *Acarapis woodi*,

Tropilaelaps spp.), nuisance caused by wax moth, ants, lizards, monkey, bears and beetles) and several diseases of fungi (*Nosema* spp., *Ascosphaera apis*), bacteria (*Paenibacillus* larvae, *Melissococcus plutonius*) and viruses. For some of these enemies and pathogens, the consequences for individual bees and colonies are known, while for others, they remain elusive.

Extinction of Insect Pollinator

Pollination is a critical ecosystem service in both managed and natural ecosystems which provide great value monetarily, ecological and others. It is the best example for mutualism of symbiosis between the pollinated crop and the pollinator. In many cases, it is the result of intricate relationships between plants and pollinators and their reduction may affect the survival of both mutualistic pollinators and their dependent plant species. Indirect evidence of pollinator loss comes from studies of pollinator communities along gradients of agricultural intensification and habitat fragmentation as proxies for temporal change (Kremen *et al.,* 2002). Most natural landscapes around the world have been disturbed by mankind activity and therefore, the pollinator abundance and richness are declining in many parts of the world. Global extinction rate of species are accelerating at an alarming rate (Table- 1). Wilson (1999) estimated that 0.2- 0.3 per cent of all species are lost every year. A range of 5-10 per cent of the tropical forest species may become extinct within the next 30 years (UNEP, 1993). The population of both wild and managed pollinators is declining at alarming rates due to change of cropping pattern, monoculture, use of agrochemicals especially during flowering period, erosion of their nesting habitats, i.e. forests and grassland ecosystems, diseases and pests, over-collecting, smuggling and trading of certain rare and endangered species.

Table 1: Estimate loss of species on earth

S. No.	Species loss	Global loss per decade (%)
1	One million species (1975-2000)	4
2	15-20% of species (1980-2000)	8
3	25% of species (1985-2015)	9

Source; Wilson, 1988

Globally, honey bee (*Apis mellifera*) is predominantly managed pollinators and play crucial role to enhance agricultural production although other species of bee e.g. the leafcutter bee *Megachile rotundata,* (Natural Research Council, 2006), bumble bee and stingless bee are also in use. In the USA, 59 per cent of honey bee colonies were lost between 1947 to 2005 and there is clear evidence for severe regional declines in domestic honey bee stocks (National Research Council, 2007; van-Engelsdorp *et al.,* 2008). There were 25 per cent of bee colonies lost from central Europe between 1985 and 2005 while agricultural crops and some wild plants dependent upon single species became worrisome (Potts *et al.,* 2010). An estimated 20 per cent of all lost honey bee colonies involved some degree of pesticide exposure. The gardeners, orchard growers, farmers and urban dwellers can switch to more pollinator-friendly organic methods of cultivation to reduce wildlife exposures to insecticides, herbicides and fungicide. Indeed, *Varroa destructor* is an ecto-parasitic mite which is an invasive species from Asia (Sammataro *et al.,* 2000). However, most of the wild and feral honey bee colonies have vanished due to mite from Europe and the USA (Moritz *et al.,* 2007; Jaffee *et al.,* 2010).

Apis cerana species is known as Indian honey bee or native bee but it remains, till now, a forgotten and completely ignored species. Therefore, from biodiversity conservation point of view, it will be disastrous to leave this important genetic resource on its own and definitely require research and development interventions for its conservation and sustainable use both in natural and agricultural eco-systems. In the Hindu Kush Himalayan range, beekeeping with *Apis cerana* is being replaced by *Apis mellifera* at such a fast pace that population of native *A. cerana* is declining to a level that is no longer viable. In Jammu and Kashmir India, 95 per cent of the beekeeping is done with exotic bee *A. mellifera* whereas *A. cerana* is confined at higher altitudes only (Abrol, 2009). In India, Nagaland is the only one state where beekeeping is flourishing exclusively with its native bee; *A. cerana*, however meliponiculture also flourish with *Tetragonula iridipennis*, while two other species of stingless bee are also perpetuated in this state; *Lophotrigona canifrons* and *Tetragonula laviceps.* Since last five years *A. mellifera* has introduced in Nagaland and there may be chance *A. cerana* vanish from Nagaland.

Until recently, there was little firm evidence of geographically widespread declines for most groups (Ghazoul, 2005). Among bees, the best documented species following the honey bee are the bumblebees (*Bombus* sp.), which have shown evidence of an ongoing decline in diversity over much of Belgium and the UK (Goulson *et al.,* 2008). There are 68 bumblebee species found in Europe and out of these, 24 per cent of European bumblebee species are threatened with extinction according to a recent study assessing the species group at the European level. The European Commission has funded two projects namely, Status and Trends of European Pollinators (STEP) and European Red List of Pollinators respectively where both projects aim to enhance pollinator population and conserve the diversity including red listed pollinators.

Butterflies are considered efficient pollinators for specialized flower bearing plant species and help to enhance the crops productivity. About 1,500 species of butterflies dwell in the Indian subcontinent, but their population is dwindling because of the indiscriminate use of insecticides and herbicides as well as atmospheric pollution. Many other manmade environmental changes affect them *viz.* deforestation, extension of farming, *Jhum* cultivation, unrestricted urbanization and monoculture. Some species of butterflies are also threatened and on the verge of extinction through destruction or disturbance of their larval as well as adult food source and shelters.

Over the past decade, farmers in the Himalayan region have been complaining about decline in apple productivity and quality due to pollination deficit caused by abated abundance and diversity of insect pollinators. The scarcity of insect pollinators causes inadequate pollination in ecosystem. The solution lies in supplementing populations of crop pollinators such as honey bees, bumblebees, stingless bees, solitary bees, etc. In USA, almond farmers and in India (Himachal Pradesh and Jammu Kashmir), apple farmers keep honey bee colonies in their orchards on rental basis to optimize the crop pollination and obtain maximum productivity.

The International Initiative for the Conservation and Sustainable Use of Pollinators was established by the 5th Conference of Parties to the Convention on Biological Diversity (CBD) as a cross cutting initiative within its work on agricultural biodiversity. FAO was invited by the Executive Secretary to "facilitate and co-ordinate the initiative in co-operation with other relevant organizations who were invited to support actions in countries subjected to pollinator decline." Awareness about the value of insect pollinators at all levels among planners, policy makers, beekeepers, farmers, students and extension workers is a must to conserve them. In western countries, farmers are already managing the crop pollination by adding additional population of honey bees, bumble bee and solitary bees (*Osmia*, *Megachille*, *Nomia*, *Xylocopa*, etc.) to optimize the pollination of different crops. The focus of beekeeping needs to change from conventional honey production to crop pollination and bee hive products.

Pollinator Deficits and Ecosystem

Globally, the declining of pollinators' population has caught the attention of scientific community. Plants suffer with pollination deficits (Kearns *et al.*, 1998). Indeed, shortages of pollinators and shortfalls of productivity are really two facets of the same problem. Long ago, Rachel Carson pointed out that silent springs and fruitless falls go hand in hand (Buchmann and Nabhan,

1996). Undoubtedly agricultural production, productivity and diversity are threatened by declining population of pollinators and their extinction. Many pollinator population densities have been reduced and the available population are unable to fulfil the required pollination service of plants; so therefore, plant production, productivity and diversity of dependent crops are in peril. In agro-ecosystem, many crops depend upon insects for their pollination and assisted pollination may have to be done when natural pollination is insufficient in order to reduce potential yield loss (Klein *et al.,* 2007). The global annual economic value of insect pollination was estimated to be $ 153 billion during 2005 (i.e. 9.5 per cent of the total economic value of world agricultural output considering only crops that are used directly for human food (Gallai *et al.,* 2009). The United Nations reported that in the economics of ecosystems and biodiversity, insect pollination was valued at £134 billion.

Some farmers are using rented insect pollinators in their orchards for fulfilment of pollination deficits to obtain the maximum productivity in the valuable cross pollinated crops which are highly dependent upon insect pollinators. A rented colony placed in a field is the primary signal of pollinator deficit; this is benefit and cost effective input but it also increases the cost of cultivation. This is mankind's big issue, sufficient pollinators and their pollination was available as free service from natural endowment was available in the ecosystem earlier whereas now, their deficit has to be fulfilled by rented pollinator colonies. Although several works have attempted to illustrate the severity of pollinator decline (Buchmann and Nabhan, 1996; Kearns *et al.,* 1998; Kevan, 1999), the problem has generally been ignored. For this reason, it is appropriate to ask the following questions from the point of view of documentation: "Are pollinators really declining?" and "Do they have economic consequences for agriculture?" We would not only answer both questions in the affirmative, but we also believe that the problem is extremely serious with far-reaching consequences for agriculture and global food production.

In Malaysia, labor costs for hand pollination are rising to fulfill the scarcity of pollinators for oil palm (*Elaeis guineensis*) Syed (1979). After careful screening and quarantine, *Elaeidobius kamerunicus* was released in Malaysian for oil palm pollination, where it rapidly became established (Syed *et al.,* 1982). They place the pollinators into a novel habitat to enhance the productivity by introduction of bumble bees into hot houses for buzz pollination of tomato in Europe (Banda and Paxton, 1991) and North America (Kevan *et al.,* 1991). Artificial pollination with electric vibrators (Kerr and Kribs, 1995) is a costly method that is no longer used, instead buzz pollination is available at lower cost and so it is favored today (Buchmann, 1983). In Canada, alfalfa crop suffered from deficit of pollination and resulted in reduction of nesting habitat; there were too few native pollinators to provide pollination except those at the vicinity of large fields (Stephens, 1955). In Manitoba, Stephens (1955) recorded yields of 1,000 kg/ha from small fields, but only 15 kg/ha from large fields. Brazil nuts are pollinated by an assemblage of large bees whose nesting habitats have been severely curtailed or even eliminated (Sutton and Collins, 1991). The habitat destruction and pollution resulting in paucity of pollinators and failures of plant reproduction and regeneration has been well described (Kevan, 1975).

The catastrophic effects of CCD on honey bee shows number of colonies are declining in North America. Simultaneously pollination service has been adversely affected, and growers have reported their grievances and difficulties regarding pollination deficit in the crops *viz.* Blueberries in Maine, Almonds in California, Cucumber in the eastern United States and hybrid seed production in western Canada. The decline of the pollinators as a result, significantly decreased the productivity of blueberry crop in the affected regions (Kevan and Plowright, 1995).

Causes of insect Pollinator Decline

There are number of causes responsible for insect pollinators decline; some causes are directly and indirectly related to mankind's which are given below:

Habitat Destruction

In developing countries, natural habitat destruction due to deforestation from a highly diverse natural ecosystems to minimize diverse ecosystem i.e., agro-ecosystems is adversely affecting native bee populations. This habitat destruction could lead to loss of different types of flowering plants and bee flora. Scarcity of bee flora not only leads to decline in colony numbers but also creates stress conditions for pollinator and increases their vulnerability to the pests and diseases. Destruction of forest habitat for growing agricultural and horticultural crops adversely affects the availability of floral resources because many of the staple crops such as rice, wheat, barley, potato, etc. are of little or no value to honey bees. It is due to increasing dearth of bee floral resources that queens would not lay eggs to increase their own population. A constant rise in human population, higher rate of deforestation, and over-exploitation of resources with expansion of farm lands for agriculture and rearing livestock cause a continuous depletion of the forest resources. The lowland and mid-hill fauna are more endangered than mountain fauna. To meet the need, the agriculture is expanding and pollinators' habitat is being lost rapidly so that sustainable agriculture may not sustain longer.

Habitat destruction is the most important factor for bee decline (Brown and Paxton 2009; Winfree *et al.,* 2009). The habitat fragmentation as a result, curtails bee diversity and abundance (Winfree *et al.,* 2007; Carre *et al.,* 2009). This might affect bee species primarily by the loss of floral and nesting resources. Habitat degradation (grazing, fire, urbanization, agricultural intensification) are few, and findings from a recent meta-analysis did not find these disturbances to have an overall significant impact on bees, although this might simply reflect low statistical power (Winfree *et al.,* 2009). However, application of herbicides eradicates the weed flora and adversely affects availability of the floral resource (Gabriel and Tscharntke, 2007; Holzschuh *et al.,* 2008). Plant biodiversity in most regions of the world has also undergone rapid change in recent decades (Lavergne *et al.,* 2006).

Injudicious use of Pesticide

Pesticides contain biologically active compounds; it is used against pest of the cultivated crops for the plant protection. Population of our country is increasing while cultivated area is decreasing, therefore, food security can be done only by virtue of enhance the productivity and protect the loss caused by insect pests. In the crop protection pesticides are used globally. Insect pollinators, beekeeping and plant protection are essential inputs for sustainable agricultural. We are not in that condition to ignore any one of them. Our increasing population has increasing food requirement, and for food security of these population each and every input play pivotal role. Since the introduction of synthetic pesticides during 1940 in the developed and developing countries, insect pollinator and beekeeping industry have been incurring heavy losses. In developed countries, large scale monoculture cultivation of crops and their plant protection pattern *viz.* exploitation phase and crisis phase create seriously amplified the problem of insect pollinator poisoning by the exposer of used pesticides against pest. Nonetheless, some area of north east in India are remaining organic state by default, whereby sustaining a great diversity of insect pollinators. However, so many efforts have been carried out towards pollinators' sustenance *viz.* training, education and arrival of new generation bio-rational molecules in the market by means minimize the loss of insect pollinators.

For pest control, a large number of pesticides are used. Farmers apply pesticides in injudicious way due to lack of knowledge about ETL, timing, suitable pesticide and their required dosage. Injudicious use of pesticides may cause severe nuisance to pollinators and disturb the natural ecosystem. Pesticide problem on pollinators is severe even in developed country like USA (loss of about $ 320 million/ year) and is equally important for other countries as well

(Abrol, 2012). Negligence of Integrated Pest Management and dependence on single control measures only with chemical methods is adversely affecting environmental health including health hazards to human beings and decline in other non-target animal populations as well as pollinators. In many parts of the world, pesticides are used to control insect pests on a large scale, whereas natural enemies and insect pollinators are usually more susceptible to the pesticide than the target insect pests. Widespread use of pesticides in many parts of the world has reduced the population and diversity of pollinators (Pimentel *et al.*, 1992) and this, particularly in the case of rare insect pollinators and rare plants, can have a devastating impact on pollination systems (Nabhan and Buchmann, 1997). There is also concern that sub-lethal doses of pesticides may disrupt the pollinating behaviour of all types of bees and render them more susceptible to diseases and parasites (Allen-Wardell, *et al.* 1998).

In order for pesticides to fulfil the role of crop protection, they must be against only targeted pests and avoid exposure of pesticides on other non-target organisms. A comprehensive body of legislation has been established to evaluate the safety of plant protection products to non-target organisms. European regulations enact to ensure that when applied properly, pesticides do not have unacceptable effects on non-target organisms such as honey bees, earthworms, fish, algae, birds, etc. Studies and risk assessments on organisms follow scientific principles found in eco-toxicology and must be completed before product registration. In European countries, pesticide should get registration in board after evaluation of no unacceptable acute or chronic effects the survival, behaviour and development, taking into account effects on pollinator, honey bees, natural enemies.

Climate Change

It is a significant and drastic change in the statistical distribution of weather patterns over periods ranging from decades to millions of years. It may be a change in average weather conditions, or in the distribution of weather around the average conditions (i.e., more or fewer extreme weather events). A number of studies suggested that climate change may be one of the biggest disturbance factors imposed on ecosystems today (Walther *et al.*, 2002 Parmesan, 2006). Observational evidence from many continents has indicated that many ecosystems are affected by regional and global climate changes, particularly temperature range increases. Those species involved in pollination interactions, this is evident through recent changes in flowering phenologies (Fitter and fitter, 2002; Miller-Rushing *et al.* 2007). The timing of pollination is determined by climatic cues such as temperature and water availability

(Cleland *et al*., 2007). Many insect pollinators also synchronies their life cycles with climatic cues, and this phenological response of plant and pollinators needs to remain broadly synchronized for many plant pollinator relationship to remain viable. Climate change is altering the phenological response of plant and some pollinators may be unable to alter their life cycles to synchronize with altered pollinating timing. Climate change, weather may not adopt the particular pollinator and sift in another region because there they may not gate suitable forage plant and climatic conditions. All these consequences may compel to extinct the insect pollinator or sift in another region.

Monoculture

Monoculture is defined as a field is cultivated of a single crop species. Monoculture can affect to duration and diversity of floral resource, resulted a short period when pollinators have more food resources than they can use, while during other periods of the year, can bring starvation. Pollen is virtually only source of protein, lipids, vitamins, and minerals for pollinators and other anthophiles. Colonies that pollinate large monocultures such as mustard have a severe lack of variability in their diets. Insect pollinators fed which nectar and pollen from diverse range of plants species had a healthier immune system than those dependent on monoculture diet. Research has shown that a diet of pure dandelion pollen will hinder larval development in mason bees, prevent brood production in honey bees, and cause 100% larval rejection in bumble bees. Monoculture increase the floral dearth period and also provide huge food availability for very short period of honey flow, but a major part of floral rewards; nectar and pollen wastage occur in nature due to flush. This is the same condition in deluge, water is essential but not in the form of flood. Monoculture creates pollination crisis by itself, this situation require a very high level of pollinators' abundance to produce economically viable crops. In most of the places, the available, pollinators do not fulfill the pollination requirement and pollination deficits happen. The increasing size of fields and orchards (monoculture) increase the importance of pollination management.

Diseases and Enemies

There are frequent reports of *A. cerana* colonies being affected by *Nosema*, virus cluster and sac brood diseases in the Hindu Kush Himalayan region. Recently, the European foul brood disease has badly affected *A. cerana* colonies in Kathmandu valley of Nepal. Amongst the mites, *Acarapis woodi*, *Varroa jacobsonii* and *Tropiolaelaps* sp. have been reported on *A. cerana*. Amongst these, Acarine disease poses a serious problem (Verma, 1987).

Amongst the predators, five different species of wasps pose a serious threat to beekeeping industry in this region. However, because of its shimmering and evasive behaviour, *Apis cerana* can resist the attacks of wasps better than *A. mellifera*. Two species of wax moths, *Gallaria mellonella* and *Archoria grisella* are serious pests in *A. cerana* colonies as this native species of honey bee do not collect propolis to guard against the attack of moths. In early eighties, the incidence and severity of Thai sac brood virus disease increased at such an alarming rate that more than 95 per cent of colonies infected in different countries were killed by this disease (Rana *et al.*, 1987). This resulted in huge economic loss to beekeeping with *A. cerana* in Asia not only in terms of honey and beeswax production but also adversely affected pollination services.

The Asian mite, *Varroa jacobsoni* is associated with *A. cerana* and *A. dorsata* bees but causes no serious problem to them, but it is fatal to *A. mellifera* colonies. The *Varroa jacobsoni* is a more serious pest of honey bees and is cosmopolitan throughout the world except a few isolated countries such as Australia. The mite feeds externally on bee larvae, pupae and adults. A regular supervision, seasonal management and preventive measures against pest should be practiced necessarily for successful beekeeping.

European honey bees have been suffering for many years from number of diseases *viz.* European foulbroods and American foulbroods, with parasites causing additional problems in recent years. The honey bee tracheal mite (*Acarapis woodi*) was discovered in the 1920s and slowly spread throughout the world, reaching the USA in the 1980s. Today it is cosmopolitan except Australia, New Zealand, Scandinavia and Canada. The mites infest the tracheal walls of young adult bees and shorten their lives, reducing honey production and pollination efficacy (Morse, 1978).

CCD (Colony Collapse Disorder)

Colony Collapse Disorder (CCD) is not a disease; it is multi-factorial syndrome that diminishes the number of adult bees in the colony which also store food supplies and bee broods. It threatens the colony by decreasing the number of adult honey bee from colony.

How to Overcome from Pollination Deficits

Public awareness

Educate the farmers, beekeepers, extension worker, agriculture officers and students about importance of pollinators in sustainable agriculture, their conservation and management. The sustenance and improvement of agriculture depends upon pollinators as well as, the effects of pollinator erosion and their deficits.

Alternative Pollinators

The extent of our reliance on single pollinator species for such an important service is risky. There is the need to identify a series of alternate pollinators' species for fulfillment of the deficit of key pollinator to sustain the crop productivity. Management of a range of bee and pollinators species is required to maintain the world's pollination systems, particularly in agricultural areas. Before management deep knowledge is essential about their taxonomy, ecology and biology.

Evaluation and Ranking of Insect Pollinator

Region wise evaluation of native pollinators in a particular crop and to catalogue them should be initiated early. Anthophiles may be categorized into two types (1) true pollinator (2) nectar and pollen robber; it can be decided on the basis of foraging attributes; scopa or other body parts carrying loose pollen tripping on the stigma or not and pollen deposited on the stigma will be categorized as true pollinators. On the basis of their foraging attributes; diurnal abundance, foraging rate, foraging speed, working period, transit time, number of loose pollen deposition on the stigma in single visit, etc. If anthophiles do not deposit pollen on the stigma those will categorized as robber of nectar/ pollen or both.

Conservation of Pollinator Dwelling and Make Buffer Zone

Buffer zones provide valuable resources for crop pollinators including shade, nesting sites, water, a variety of nectar and pollen free from pesticides. Buffer zones can reduce wind and enhance the foraging and pollination efficiency. Ideally, buffers should be < 1,000 feet from the cultivated crop. Farmers might be encouraged to protect "corridors" that connect natural habitats, or uncultivated areas within and around cultivated ones.

Adopt to Integrated Pest Management

Utilize all suitable techniques for the pest control in a compatible manner to suppress the pest population below ETL. Use of the non chemical methods for pest management at blooming stage of crop should be encouraged and if it is necessary, use safer and selective pesticides with sticking agent at the evening time (after cessation of pollinator foraging).

Before spraying of chemicals use conscience and take preventive measures given below:

1. Restrict pesticide applications to the evening only in order to avoid the toxic exposure during foraging period of insect pollinators.
2. Choose the selective, high dissipation, highly degradable and safer pesticides.
3. Avoid chemical spray on plants during flowering stage.
4. Limit the application rate of plant protection products.
5. Use hood for reducing the drift on neighbouring flowering crops.
6. Remove flowering weeds from vegetative cropped areas prior to chemical application.

Conservation and Utilization of Pollinators

For several decades, researchers have been trying to conserve pollinators like honey bees providing nesting sites and quality of forage, and protecting them from pesticides. Managed pollination in crops has been a largely neglected part in agriculture, which requires attention to increase productivity and quality. In the developed countries, rented insect pollination has increased considerably during the past few decades and arrangements for insect pollination are now a part of standard management practices when growing many crops. In USA, Canada, Germany, etc. millions of honey bee colonies are being hired on rental basis for pollination services in valuable commercial crops. In India, it is also adopted in apple orchards in Himachal and Jammu Kashmir.

As far as conservationists are concerned, most emphasis has been given to large mammals, birds, and reptiles but very less importance has been given to insect pollinators except honey bees. Biodiversity cannot be isolated from pollinators' diversity, and therefore, there is a need to address pollinators and their conservation issues in existing acts and regulations to take care of pollinator issue. Community members as users of local resources should be aware about importance of wild bee conservation for environment improvement and benefit sharing. More research converges to conserve insect pollinators, honey bees and other natural pollinators to explore their credential pollination potentiality.

National Policy for Pollinators

Indubitably, insects are the key pollinating agents for many agricultural and horticultural crop and other plants. Their role of pollination for sustainable agriculture to sustain the productivity and maintaining biological diversity is beyond the imagination. In India, till now, beekeeping is practiced mostly for production of bee hive products specially honey and wax whereas honey bee pollination service has been providing 13- 20 times higher economic return in comparison to bee hives products.

International Policy for Pollinators

The (CBD) Convention on Biological Diversity decision in its third meeting of the Conference of the Parties recognized the importance of agricultural biodiversity and decided to establish multiyear activities on Agro-bio diversity. It was decided to take attention on priority of the components of biological diversity responsible for the maintenance of ecosystem services important for the sustainability of agriculture, including pollinators. It was decided to encourage the interested international organizations to conduct case studies on pollinators, monitoring of the loss of pollinators worldwide; the identification of the specific causes of pollinator decline; the estimation of the economic cost associated with reduced pollination of crops; the identification and promotion of best practices and technologies for more sustainable agriculture; and the identification and encouragement of the adoption of conservation practices to maintain pollinators or to promote their re-establishment.

The Fifth Conference of Parties in 2000, an International Initiative for the Conservation and Sustainable Use of Pollinators (also known as the International Pollinator Initiative - IPI) was established and requested the development of a plan of action. In April 2002, the Sixth Conference of Parties established the Plan of Action of the IPI. The Plan of Action of the IPI consists of four basic elements: (i) assessment (ii) adaptive management (iii) capacity building and (iv) mainstreaming. Through its Global Action on Pollination Services for Sustainable Agriculture, AGP has been coordinating and facilitating the implementation of the IPI by undertaking, in collaboration with numerous partners, activities that contribute to the implementation of these four elements, and to address the management of pollination services for sustainable agriculture. The IPI have objectives to promote coordinated action on world-wide pollinator decline, its causes and its impact on pollination services, address the lack of taxonomic information on pollinators; assess the economic value of pollination and the economic impact of the decline of pollination services; and to promote the conservation and the restoration and sustainable use of pollinator diversity agriculture and related ecosystems.

A number of other activities and initiatives have been developed and are being implemented, to respond to the issues related to the conservation and sustainable use of pollinators. A few examples of such initiatives include the African Pollinator Initiative, The Brazilian Pollinator Initiative, the European Pollinator Initiative and the North American Pollinator Protection Campaign, etc. The International Centre for Integrated Mountain Development (ICIMOD) is working for pollinator conservation. India is also undertaking extensive work in the area of pollination in the Hindu- Kush Himalaya region.

References

Abrol, D.P., 2012. Pollination biology- biodiversity conservation and agricultural productivity. Springer Dordrecht Heidelberg London New York, 787 pp.

Abrol, D.P. 2009. Bees and beekeeping in India, 2nd edn. Kalyani Publishers, Ludhiana, 720 p.

Aizen, M. and Harder, L.D. 2009. The global stock of domesticated honey bees is growing slower than agricultural demand for pollination. Current Biology, 19:915-918.

Allen-Wardell, G., Bernhardt, P., Bitner, R., Burquez, A., Buchmann, S., Cane, J., Cox, P.A., Dalton, V., Feinsinger, P., Ingram, M., Inouye, D., Jones, C.E., Kennedy, K., Kevan, P., Koopowitz, H., Medelllin, R., Medellin-Morales, S., Nabhan, G.P., Pavlik, B., Tepedino, V., Torchio, P. and Walker, S. 1998. The potential consequences of pollinator declines on the conservation of biodiversity and stability of food crop yields. Conservation Biology, 12(1):8-17.

Banda, H.J. and Paxton, R.J. 1991. Pollination of greenhouse tomatoes by bees. Acta Horticulture, 288:194-198.

Biesmeijer, J.C., Roberts, S.P. M., Reemer, M., Ohlemuller, R., Edwards, M., Peeters, T., Schaffers, A.P., Potts, S.G., Kleukers, R., Thomas, C.D., Settele, J. and Kunin, W.E. 2006. Parallel declines in pollinators and insect-pollinated plants in Britain and the Netherlands. Science, 313:351-354.

Brown, M.J.F. and Paxton, R.J. 2009. The conservation of bees: a global perspective. Apidologie, 40:410-416.

Buchmann, S.E. 1983. Buzz pollination in angiosperms. Handbook of experimental pollination biology. Van Nostrand Reinhold, New York, pp 73-113.

Buchmann, S.E. and Nabhan, G.P. 1996. The forgotten pollinators. Island Press, Washington, DC.

Carre, G., Roche, P., Chifflet, R., Morison, N., Bommarco, R., Harrison-Cripps, J., Krewenka, K., Potts, S.G., Roberts, S. P.M. Rodet, G., Settele, J., Steffan-Dewenter, , I., Szentgyorgyi, H., Tscheulin, T., Westphal, C., Woyciechowski, M. and Vaissiere, B.E. 2009. Landscape context and habitat type as drivers of bee diversity in European annual crops. Agriculture Ecosystem Environment, 133: 40- 47.

Daily, G. C., Alexander, S., Ehrlich, P.R., Goulder, L., Lubchenco, J., Matson, P.A., Mooney, H.A., Postel, S., Schneider, S.H., Tilman, D. and Woodwell, G.M. 1997. Ecosystem services: benefits supplied to human societies by natural ecosystems. Issues in Ecology: 1-16.

FAO. 2007. Pollinators: Neglected biodiversity of importance to food and agriculture CGRFA-11/07/Inf.15. Food and Agriculture Organization of United Nations, Rome.

Gabriel, D. and Tscharntke, T. 2007. Insect pollinated plants benefit from organic farming. Agriculture Ecosystem Environment, 118: 43- 48.

Gallai, N., Vaissie`re, B. E., Potts, S. G. and Salles, J. 2009. Assessing the monetary value of global crop pollination services. Oxford University Press.

Ghazoul, J. 2005. Buzziness as usual? Questioning the global pollination crisis. Trends of Ecology Evolution, 20:367-373.

Goulson, D., Lye, G.C. and Darvill, B. 2008. Decline and conservation of bumble bees. Annual Review Entomology, 53:191-208.

Holzschuh, A., Steffan-Dewenter, I. and Tscharntke. T. 2008. Agricultural landscapes with organic crops support higher pollinator diversity. Oikos, 117:354-361.

Jaffé, R., Dietemann, V., Allsopp, M.H., Costa, C., Crewe, R.M., Dall'Olio, R., De la Rúa, P., El-Niweiri, M.A.A., Fries, I., Kenzic, N., Meusel, M.S., Paxton, R.J., Shaibi, T., Stolle, E. & Moritz, R.F.A. 2010. Estimating the density of honey bee colonies across their natural range to fill the gap in pollinator decline censuses. Conservation Biology, 24:583-593

Kearns, C.A., Inouye, D.W. and Waser, N.M. 1998. Endangered mutualisms: the conservation of plant- pollinator interactions. Annual Review Ecology Systematic, 29:83-112.

Kerr, E.A. and Kribs, W. 1955. Electric vibrator as an aid in greenhouse tomato production. Queensland Journal of Agriculture Science, 2:157-169.

Kevan, P.G., Straver, W.A., Offer, M. and Laverty, T.M. 1991. Pollination of greenhouse tomatoes by bumblebees in Ontario. Proceeding of Entomology Society, Ontario 122:15-19.

Kevan, P.G. 1975. Pollination and environmental conservation. Environment Conservation, 2:293-298.

Kevan, P.G. 1999. Pollinators as bioindicators of the state of the environment: species, activity and diversity. Agriculture Ecosystem Environment, 74:373-393.

Kevan, P.G. and Plowright, R.C. 1995. Impact of pesticides on forest pollination. In: Armstrong JA, Ives WGH (eds) Forest insect pests in Canada. pp 607-618.

Klein, A-M., Vaissi-Are, B. E., Cane, J.H., Steffan-Dewenter, I., Cunningham, S.A, Kremen, C. and Tscharntke, T. 2007. Importance of pollinators in changing landscapes for world crops. Proceedings of the Royal Society B: Biological Sciences 274: 303-313.

Kremen, C., Williams, N. and Thorp, R. 2002. Crop pollination from native bees at risk from agricultural intensification. Proceeding National Academic Science, USA 99:16812-16816.

Kremen, C., Williams, N.M., Aizen, M.A., Gemmill-Herren, B., LeBuhn, G., Minckley, R., Packer, L., Potts, S.G., Roulston, T., Steffan-Dewenter, I., Vazquez, D.P., Winfree, R., Adams, L., Crone, E.E., Greenleaf, S.S., Keitt, T.H., Klein, A-M, Regetz, J. and Ricketts, T.H. 2007. Pollination and other ecosystem services produced by mobile organisms: a conceptual framework for the effects of land-use change. Ecology Letter, 10:299-314.

Lavergne, S. et al. 2006. Fingerprints of environmental change on the rare Mediterranean flora: a 115-year study. Global Change Biology, 12:1466-1478.

Moritz R.F.A., Kraus F.B., Kryger P., Crewe R.M. 2007. The size of wild honey bee populations (Apis mellifera) and its implications for the conservation of honey bees. Journal of Insect Conservation, 11:391-397.

Morse, R.A. 1978. Honey bee pests, predators and diseases. Cornell University Press, Ithaca, 430 pp.

Nabhan, G. P. and Buchmann, S. 1997. Services provided by pollinators. In G. C. Daily (Ed.), Nature's services: Societal dependence on natural ecosystems (133-150). Washington, DC: Island Press.

National Research Council. 2007. Status of pollinators in North America. National Academies Press, Washington, DC: 307 p.

Natural Research Council. 2006. Status of pollinators in North America. National Academic Press, Washington, DC.

Pimentel, D., Acquay, H., Biltonen, M., Rice, P., Silva, M., Nelson, J., Lipner, V., Giordano, S., Horowitz, A., and D'Amore, M., 1992, Environmental and economic costs of pesticide use, BioScience 42:750-760.

Potts, S.G., Roberts, S.P.M.., Dean, R., Marris, G., Brown, M.A., Jones, H.R., Neumann, P. and Settele, J. 2010. Declines of managed honey bees and beekeepers. European Journal of Apicultural Research, 49(1):15-22.

Rana, B.S., Garg, Paul, I.D., Khurana, S.M., Verma, L.R. and Agrawal, H.O. 1987. Thai sacbrood virus of honey bees (Apis cerana indica F) in the North West Himalayas. Indian Journal of Virology, 2:127-131.

Sammataro, D. Gerson, U. and Needham, G.R. 2000. Parasitic mites of honey bees: life history, implications, and impact. Annual Review Entomology, 45:519-548.

Sciencedaily. 2008. Economic Value Of Insect Pollination Worldwide Estimated At U.S. $217 Billion.

Steffan-Dewenter, I., Potts, S.G. and Packer, L. 2005. Pollinator diversity and crop pollination services are at risk. Trends of Ecology Evolution, 20:651-652.

Stephen, W.P. 1955. Alfalfa pollination in Manitoba. Journal of Economic Entomology, 48:543-548.

Sutton, S.L. and Collins, N.M. 1991. Insects and tropical forest conservation. In: Collins NM, Thomas JA (eds) The conservation of insects and their habitats. Academic, London, pp 405-422.

Syed, R.A., Law, I.H. and Corley, R.H.V. 1982. Insect pollination of oil palm: introduction, establishment and pollinating efficiency of Elaeidobius kamerunicus in Malaysia. Planter (Malaysia), 58:547-561.

Telegraph, One third of Europe's butterflies in decline, according to Red List, 2010.

UNEP. 1993. Global biodiversity. UNEP, Nerobi.

Van-Engelsdorp, D., Hayes Jr., J., Underwood, R.M., Pettis, J. 2008. A survey of honey bee colony losses in the U.S., Fall 2007 to Spring 2008. PLoS ONE 3:e4071. doi:DOI:10.1371/journal.pone.0004071

van-Engelsdorp, D., Underwood, R., Caron, D., Hayes, J. Jr. 2007. An estimate of managed colony losses in the winter of 2006-2007: a report commissioned by the apiary inspectors of America. American Bee Journal, 147:599-603.

Verma, L.R. 1987. Pollination ecology of apple orchids by hymenopterous insects in Matiama Narkanda temperate zone. Final report. Ministry of Environment and Forests, Govt. of India, p 118.

Wilson, E.O. 1988. The current state of biodiversity. In: Wilson EO, Peter FM (eds) Biodiversity. National Academic Press, Washington DC, pp 3-18

Wilson, E.O. 1999. The diversity of life, new editionth edn. W.W. Norton and Company, Inc., New York

Winfree, R., Aguilar, R., Vázquez, D. P., LeBuhn, G. and Aizen, M. A. 2009. A meta-analysis of bees' responses to anthropogenic disturbance. Ecology, 90:2068-2076.

Winfree, R., Griswold, T. and Kremen, C. 2007. Effect of human disturbance on bee communities in a forested ecosystem. Conservation of Biology, 21:213-223.

4

CCD; Causes and Their Preventive Measures

Global Prospect of CCD

The Colony Collapse Disorder (CCD) is not a disease; it is multi-factorial syndrome which causes massive losses of honey bee in the USA and elsewhere. It is recorded that in European honey bee (*Apis mellifera*). The adult worker bee population of the colony demean through catastrophic effects of CCD. The dwindling of adult worker bees in colonies and the lack of noticeable symptoms leads to the nebulous label of CCD. In this syndrome, dead worker bees cannot be found in the colony and their vicinity of bee hive, and stored food (honey and pollen) are not touched by robbers bees and combs remain as such for several weeks even after the collapse. Some scientists; Aizen and Harder, also reported this disorder is not universal and, in fact, global populations of honey bee (*A. mellifera*) populations are increasing. In the spring of 2007, beekeepers in the US reported a rare kind of bee colony disease where they die off and numerous colonies left their hives and disappeared without a trace. The plight of honey bee in USA has caused concern and attracted great attention of scientists to delve and provide explanations of this mystery and preventive measures. In US, it is also christened "Honey bee AIDS" by some popular media. Indeed, honey bee colonies can die in many ways and CCD is just one of them. Finally, since both honey bee host and pathogens are genetically diverse, the symptoms and causes of colony losses may well be different in varrious regions.

CCD (Colony Collapse Disorder)

Colony Collapse Disorder (CCD) is not a disease; it is multi-factorial syndrome which leads to diminished number of adult worker bees in the colony while stored food and bee broods remained in the colonies. It threatens the colony by decreasing the number of adult honey bee from the colony.

The following symptoms may be observed stepwise:

1. The queen is present in colony with insufficient workers and, consisting mostly of young adult bees.
2. Presence of capped brood.
3. Presence of food stores; honey and pollen.
4. Eventually, complete absence of adult bees in colonies, with none or little build up of carcass in the colonies.
5. Adult worker bees carcass do not present in the colony and nearby the beehive.

Causes of CCD

The causes of CCD are not yet known. The potential causes investigated by "The CCD working group" are given below:

1. **Parasites and Pathogens-** Varroa mite is the most serious enemy of the honey bee; this mite is a parasite and also act as a vector because they transmit viruses. The chemicals used to control varroa mites further compromise the honey bees' health and quality of bee hive products. The key pests of honey bees, such as American foulbrood and tracheal mites, do not lead to Colony Collapse Disorder on their own but some suspect they may make bees more vulnerable and susceptible to it. The *Nosema ceranae* was present in the digestive tracts of some colonies with symptoms of CCD. Adult and broods of bees have been suffering from numbers of disease *viz.* European foul brood (caused by *Mellisococcus pluton*), American foul brood (caused by *Paenibacillus larvae*) on larvae and pupae.
2. **Remedy residue and their contamination-** Beekeepers have been using chemical or mixture of chemical at par with the crisis phase of colonies to control mites and diseases of bees. It is due to evolution of resistances that pests may not be controlled on recommended dosage. The chemical residue present in stored food and wax, which are inimical to worker bees further leads to curtail their longevity.
3. **Malnutrition-** Wild honey bees forage on the divers of flowers in their habitat and enjoy a variety of pollen and nectar sources. In commercial beekeeping, rented honey bee colonies are used in limited foraging area on specific crops of their dwelling vicinity. Colonies kept by beekeepers in suburban and urban neighborhoods offer limited plant diversity. Honey bees which forage on single crops or limited varieties of plants

may suffer nutritional deficiencies that stress their immune systems. Malnutrition does affect the fitness of the adult bees, their activity and queen fecundity as well as ultimately, the colony strength.

4. **Electromagnetic radiation-** A widely-reported theory that cell phones may be playing a role in Colony Collapse Disorder proved to be an inaccurate representation of a research study conducted in Germany. Scientists looked for a link between honey bee behavior and close-range electromagnetic fields. They concluded that there is no correlation between the inability of bees to return to their hives and exposure to such radio frequencies. The scientists strongly disavowed any suggestion that cell phones or cell towers are responsible for CCD.

5. **Level of stress-** Commercial beekeepers rent their hives to farmers. These bee hives are carried on tractor trailers, covered, and driven for hundreds or thousands of miles in covered and caged condition. As such bees which are relocated every few months must be under stressful conditions. In adult bees, analysis indicated by stress induced proteins symptoms which may also affect their behaviour and life.

6. **Inbreeding depression-** Beekeepers usually start beekeeping with a limited number of bee colonies and multiply them from same mother stock which is repeatedly multiplied from same mother stock. Therefore, bees of this group have less genetic diversity and they may suffer inbreeding depression which is well known and common in those insects of 4- 5 generation multiplied by same mother stocks. These bees become more vulnerable against diseases. The limited genetic pool may degrade the quality of queen bees used to start new hives, and the resulting progeny of these honey bees are significantly more susceptible to diseases and pests.

7. **Pesticides-** Most of the world's agro-ecosystems pest management is dependent on pesticides whereas ideal agro-ecosystems have been adopting Integrated Pest Management system and others have been adopting exploitation phase, crisis phase and some disaster phase also of the chemical control. Injudicious use of pesticides at blooming stage adversely affects bee foraging, causes heavy mortality to worker bees due to exposure of pesticides. Bee mortality are responsible for shortage of colony resulted pollination deficits, food security and toxicated nectar and pollen. Beekeepers are particularly concerned about a possible connection between Colony Collapse Disorder and neonicotinoids. It is corroborated because imidacloprid mode of action affects similar to the symptoms of CCD. Identification of a causative pesticide will likely

require studies of pesticide residues in the honey or pollen abandoned by affected colonies.

8. **Monoculture-** A monoculture is defined as a field composed of a single crop species. Scanty studies were accomplished on the detrimental effects of monocultures on the health of honey bee. Bees fed with pollen from diverse range of plants species had a healthier immune system than those dependent on monoculture diet. Diversity in diet enabled bees to protect themselves and their larvae from microbes and pathogens more effectively. Monoculture provides only one type of pollen as a food source and can lead to certain nutrient deficiencies. Managed field margins offer a means of reducing the impact of agricultural monocultures within intensively managed environments. Monoculture leads to long floral dearth for honey bee and it also affects the population density.

9. **Introduction of exotic pollinator-** Introduction of exotic species for crop pollination and native pollinators through inters competition for resources. During honey flow period, food competition does not affect them whereas during floral dearth period internal food competition becomes inimical for their survival, growth and multiplication. However, it still remains controversial whether competition actually occurs and impacts native bee population viability. In India, the number of *Apis cerana* colonies have been decreasing since the introduction of *A. mellifera* and this observation is clearly reflected in *A. mellifera* dominated beekeeping areas *viz.* Punjab, Haryana, Bihar, Himachal, U. P., etc. whereas in Nagaland, beekeeping deals exclusively with native bee species, *A. cerana.* Introduction of exotic bees in new areas may introduce their diseases and natural enemies which may later infect and infest other related species. There is clear evidence that translocated alien bees can increase the risk of spreading harmful pathogen.

10. **Genetically Modified Crops** – There is suspicion on the pollen of genetically modified crops, specifically, corn altered to produce Bt (*Bacillus thuringiensis*) as a possible source of some Bt toxin. Most researchers agree that exposure to Bt pollen alone is not a likely cause of Colony Collapse Disorder. The transgenic crops developed by genetic engineering as Bt. crops causes *B. thuringiensis* effects on insect feeder/ forager insects. The practice of incorporating the Bt. crops has raised concerns about the effect of pollen from these plants on to pollinators, but so far the evidence is scant. German researchers noted a possible correlation between exposure to Bt pollen and compromised immunity to the fungus *Nosema*.

Recommendations and Preventive Measures

Best Recommendation against CCD for Beekeepers -Since little is known for sure about the cause(s) of CCD, mitigation must be based on improving general honey bee health and habitat and countering known mortality factors by using best management practices. This includes supplemental feeding in times of nectar/pollen scarcity.

Best Recommendations against CCD for the Public- The public can take measures to improve honey bee survival with judicious use of pesticides. In particular, the public should avoid applying pesticides at blooming stage, and if it is necessary, to avoid during mid-day hours when honey bees foraging is at its peak. It may be permissible during evening at cessation of bee foraging. In addition, the public can plant pollinator-friendly floras; those that produce good sources of nectar and pollen.

Preventive Measures

1. Avoid pesticide sprays during flowering period, and if necessary, use of bio-rational, selective and safer pesticides during evening after cessation of bee foraging.
2. In order to avoid homozygous population and inbreeding depression, replace the queen with another which is taken from other apiary for genetic conservation and prevent the erosion of genetic pool.
3. Ensure the colony is strong and hygienic.
4. Use of scientific, recommended chemicals and their dosage for control of diseases and enemies.
5. Provide diversity of bee flora to provide variety of pollen and nectar sources the whole year which will avoid nutritional deficiencies and enhance their immune systems.
6. Keep bee colonies away from electro high magnetic areas.
7. Avoid frequent and long transportation to reduce the level of stress.
8. Study of comprehensive assessment of the ecosystem risk before introduction of exotic pollinators.

5

Management of Pollinators

Pollination is a keystone process in both human managed and natural terrestrial ecosystems. Pollinators are an input of crop cultivation and associated biodiversity, and provide an essential ecosystem service to both natural and agro- ecosystems. In agro-ecosystems, management of pollinators can maximize the pollination and thereby enhance the crop productivity and yield quality. Pollination is the best example of mutualism between plant species and their pollinators which results into intricate relationships between plant and animal - the reduction or loss of either affecting the survival of both.

Approximately 80 per cent of all flowering plant species are dedicated to pollination by animals and mostly by insects. The negative impact of pollinator deficit is strongly felt in reduction of fruit set seed productivity and agricultural biodiversity. The role of pollinators is transfer of pollen from anther to stigma of the same species to ensure reproduction, seed set, fruit set development and multiplication of plants, both in agro ecosystems and natural ecosystems. Indeed, some plant species rely upon only few specialized pollinators to provide pollination services and their population also affects productivity and plant biodiversity. Indigenous species particularly, have been subjected to threatened by such as natural habitat destruction and fragmentation for agricultural purposes, monoculture, pesticide use, tourism, introduction of exotic species, etc. Current understanding of the pollination process shows that, while interesting specialized relationships exist between plants and their pollinators, healthy pollination services may ensure an abundance and diverse pollinators groups.

How to Manage Pollination Services

Bees are essential components of almost all of the world's terrestrial ecosystems. They provide both pollination services and are excellent indicators of the terrestrial environments including responses to global warming. There is a major challenge to learn and understand the management of wild bees and their pollination services. There is also the taxonomic impediment in identifying the exact name of pollinators. The taxonomic identification of bee assists in knowing many aspects of its ecology and natural history - where it nests, for instance, and how it may be affected by climate. Bee experts are too few to provide identification services all around the world. Furthermore, a beginner trying to identify bees faces big challenge.

Pollinator Management at Different Levels

FAO, working with experts, has carried out an initial survey of existing good practices to conserve the managed and wild pollination services, upon which new practices can be based. Farmers can introduce measures to promote pollinators at a variety of different scales. Most commonly, farmers can make modifications on how they apply farming practices in their fields that will benefit pollinators. But there are other scales of management that are also important, from practices on non-farmed land and field edges to management of agricultural landscapes.

Field Level

At the field level, pollinator-friendly practices include minimizing the use of farm chemicals through organic production, integrated pest management or alternatives to bio-rational agrochemicals. A reduction in the use of herbicides as well as pesticides creates congenial ambiance for augmentation and conservation of pollinators' colonies in the crop fields and orchards. Cardamom farmers in the Western Ghats, India, are learning to ingeniously manipulate shade-tree cultivation in their fields to ensure continuity of pollinators. Because cardamom requires pollinators for fruit production, it is crucial to ensure that large numbers of pollinators are available during the blooming season. Most pollinators of cardamom are wild and thus move freely through the landscape. Many cardamom farmers also cultivate coffee which has an even shorter mass blooming season. Farmers are maintaining to plant a diversity and stagger planting to get continuous of flowering tree species called "sequential blooms" that provide reliable pollen and nectar resources for valuable native bees at certain times of the year when neither cardamom nor coffee is blooming. Apple growers in Kashmir and Himanchal Pradesh hire rented bee colonies during blooming period to maximize the pollination.

Farm Level

The way farmers organise different land uses across their farms can influence pollination services. In Colombia, farmers recognised that they encourage pollinator populations by conserving diverse cropping patterns in their farms, for example, by combining mixed cropping, kitchen gardens and agro forestry systems, as also providing habitat in their farms for pollinators. Similarly, farmers in Tanzania understood and encouraged the nesting of carpenter bees in their houses despite some minor structural damage. It was very unfortunate and I pity when, in Ruzaphema village of Nagaland, I saw children were killing carpenter bees. On my enquiry as to why they were doing that, they told it is a pest of wood which damage the wooden planks of thatch roof. It is the fault of our and the government policies because we have not educated them till now that it is not a pest but that it is a beneficial insect and works as a pollinator.

Landscape Level

Farmers from many regions recognised that they benefited from areas of natural vegetation in close proximity to farmland. Such habitat patches provided flowering resources and nesting sites that sustain pollinators. Practical applications of good practices in on-the-ground settings are informative for people learning to better manage pollination services. Evaluation of good practices for their impacts on pollinators and their relative costs and benefits to farmers and land managers is also useful, since the value of these practices must withstand the test of providing sufficient benefits, considering the time, effort and costs of implementing them.

In multiple agro-ecosystems and ecologies, pollinator-friendly management practices have been identified that serve to enhance yields, quality, diversity and resilience of cropping systems:

- Conserve wild habitat.
- Manage cropping systems, flower-rich field margins, buffer zones and ensure habitat and forage.
- Grow shade trees and provide bee nest sites, e.g. by leaving standing dead trees and fallen branches undisturbed.
- Reduce the application of pesticides and associated risks or manipulated the time of application.
- Establish landscape configurations that favour pollination services.
- Provide nesting materials on the farms.
- Grow bee flora for continuous provide feed to the pollinators.

Pollination Information Management System (PIMS)

The (PIMS) is being developed through collaboration between FAO and a number of national level organizations concerned with the conservation and sustainable use of pollination services for sustainable agriculture. The PIMS serves to organize and deliver accurate information on managing pollination services of key crops, globally, to farmers, farm advisors and land managers. The system is designed to help users answer the following questions:

- ➢ What are the pollination needs of a particular crop?
- ➢ What is the current understanding of managing the pollination of a particular crop?
- ➢ What studies have been carried out on the pollination of the crop?
- ➢ What is known about the pollinators of the crop?

Augmentation of Pollinators

Augmentation is an effort to increase population of insect pollinators either by proliferation or by environmental manipulation. Augmentation is the placing of additional population among the naturally occurring pollinators in the ecosystem to fulfill the deficits of pollination. Its objectives may be achieved by proliferation of pollinators in the ecosystem or by modification into congenial environment to promote their population abundance and their effectiveness. This concept may fulfill the pollinator deficits in ecosystem and increase the crop productivity fulfillment of proliferated population food security. It also helps to sustain the pollinator diversity, thereby conserve and maintain the plant diversity. Augmentation may cause inbreeding depression; the augmented populations with rich diverse populations have higher diversity or added natural population for propagation and it may be more successful with this approach. Maximum production and productivity of apple fruit was obtained through natural pollination with augmentation of honey bees (Khan *et al.,* 2012). In California, about 1.4 million bee hives are rented annually to augment natural pollination of almonds (about 50 per cent of the hives), alfalfa, melons, and other fruits and vegetables (Pimentel *et al.,* 1992b).

The author envisages about some exciting research which have been going on in U. S. Department of Agriculture's Specialty Crop Research Initiative (SCRI). The Project Team members are investigating the performance, economics, and farmer's perceptions of different pollination strategies in various fruit and vegetable crops, habitat manipulation to provide congenial ambiance for bees and use of managed honey bees alone or in combination with wild bees. Its long-term goal is to develop and deliver context-specific Integrated Crop Pollination (ICP) recommendations on how to most effectively harness the potential of native bees for crop pollination. ICP may be defined as the combined use of different pollinator species, habitat augmentation and crop management practices to provide reliable and economical pollination of crops.

This approach is analogous to Integrated Pest Management to provide decision-support tools to reduce risk and improve returns through the use of multiple tactics tailored to specific crops and situations. It is by developing context-specific ICP programs that may improve sustainability of specialty crops and thereby, help ensure the continued ability of growers to reap profitable returns from their investments in land, plants, and other production inputs.

The Specialty Crop Research Initiative (SCRI) has five objectives:

1. To identify economically valuable pollinators and the factors affecting their abundance.
2. To develop habitat management practices to improve crop pollination.
3. To determine performance of alternatively managed bees as specialty crop pollinators.
4. To demonstrate and deliver ICP practices for specialty crops.
5. To determine optimal methods for ICP information delivery and measure ICP adoption.

Conservation Measures

Conservation is the set of actions to protect and increase pollinators' population density by manipulation of environment and their activities. Pollinators' conservation refers to the enhancement of naturally occurring pollinators' density as part of crop pollination strategy. Depleting of concerned pollinators' population and diversity in agro-ecosystems and natural-ecosystem are critical to sustain productivity and plant biodiversity. Loss of native pollinators due to their habitat loss (deforestation, eradication of natural ecosystem, heavy tillage), diseases and natural enemies, injudicious use of pesticides, monoculture (enhance the floral dearth period), climate change and many more. Buchmann and Nabhan, 1996 have reported depletion of pollinators in ecosystem.

Assessment of Pollinators

Studies of pollinators' assessment, status and trends are intricate and it is to catalogue the true insect pollinators' area wise and crop wise. Thereafter, evaluation of catalogued pollinators' on the basis of their credential pollinating abilities and their rankings is most important for pollination management. Assessments of pollinator crisis and pollination deficits may explicitly curtail productivity and plant biodiversity. Such types of studies are needed to make a comprehensive analysis of the status and trends of pollinators to collect useful information for enhancing pollinator conservation and sustainable use. Measurement and assessment of pollination services therefore, needs to be properly planned and designed to produce verifiable results. To accurately assess the extent of pollinator declines, standardized methodologies should be applied globally. There are methods in existence to collect information for assessing the monetary value of pollination services for commercial crops, but more collaborative studies between agricultural economists, natural resource and environmental economists, agronomists and pollination biologists should be fostered.

Management of Pollination Services

Pollinators have been playing pivotal role in pollination services for seed production as well as enhancing the productivity, quality and maintaining the biodiversity. In spite of precious service, these areas are neglected by the governments, policy makers and scientists also. In sustainable agriculture, in order to sustain the productivity and maintain the biodiversity, policymakers should frame policies with explicit mandates to conserve and proficiently manage pollination services. Strategic ways are needed to mainstream pollination concerns into the relevant sectors and promote "pro-pollinator" actions.

International Initiatives

In recognition of alarming pollination crisis, there has been a mobilization of effort on several levels to address pollination management and conservation. On a global level, the international community has identified the importance of pollinators and considering the urgent need to address the issue of the worldwide decline in pollinator diversity, the Conference of the Parties to the Convention Biological Diversity established an International Initiative for the Conservation and Sustainable Use of Pollinators. In November 2000, FAO organized a meeting with the participation of key experts to discuss on how to elaborate the International Pollinators Initiative. Subsequently, a plan of action was prepared by FAO and the CBD Secretariat. The aim of the International Initiative for the Conservation and Sustainable Use of Pollinators (IPI) is to promote coordinated action worldwide: to monitor pollinator decline, its causes and its impact on pollination services; to address the lack of taxonomic information on pollinators; Assess the economic value of pollination and the economic impact of the decline of pollination services; and, promote the conservation, restoration and sustainable use of pollinator diversity in agriculture and related ecosystems.

National Initiatives

The National Commission on Agriculture in 1976, recommended the starting of a project entitled "All India Coordinated project on Honey bee Research & Training", which was launched by the Indian Council of Agricultural Research in 1980- 81 with 6 centers in India. By the time importance of pollinators in sustainable agriculture was realized by scientists and policy makers of India, developed countries had realized pollinators were declining where many versed research workers had published so many papers. While honey bee pollination is 13- 18 times more lucrative than bee hive product, 80 per cent crops rely on pollinators for their pollination and many more aspects. The AICRP (Honey Bees & Pollinators) objectives were refined and the title of this project was re-designated as All India Coordinated Research Project on Honey Bees & Pollinators in July 2007. This project is has 23 centers across India and expansion is continued till date.

In India; the Ministry of Environment and Forest, Government of India, identified the G.B. Pant Institute of Himalayan Environment and Development (GBPIHED), as the national executing agency for the project on "Conservation and Management of pollinators for sustainable Agriculture, through ecosystem approach" that is being implemented by UNEP, approved by GEF and executed by FAO. The development objective of the project is improved food security, nutrition and livelihoods through enhanced conservation and sustainable use of pollinators. The outcome of the project will be an expanded knowledge of pollination services, conservation and sustainable use of pollinators for sustainable agriculture. China has officially recognized pollination as an agricultural input, along with other conventional inputs such as fertilizers and pesticides. But the degree to which pollination can contribute to sustainable crop yields has not been addressed in agricultural policies in most countries.

The Centre for Pollination Studies (CPS) has been established at the University of Calcutta through the Darwin Initiative of the Department for Food and Rural Affairs (DEFRA) in partnership with the Government of India, Department of Science & Technology. The CPS is focused on pollination systems with an expectation that as the centre grows; the research will extend into other areas that contribute to the development of sustainable farming.

There are three principal aims:

- To increase understanding of native pollinators.
- To improve the management of pollinators and the wider agro-ecosystem.
- To improve the livelihood of subsisting farmers and contribute to a healthy agro-ecosystem through the above routes.

CPS has begun work in two regions of northeast India, and in each it has established regional field study and advice centres.

Initiative of AICRP (Honey Bees & Pollinators)

G. B. Pant University has protected an area under AICRP (Honey Bees & Pollinators) in 2013 for conservation of natural flora and fauna biodiversity as a bee pasture. The European community has identified and protected 431 Prime Butterfly Areas in 37 sites where natural sites are protected because they serve as habitat. It has been noted that for insect pollinators, the design of protected areas will require special considerations. There are multiple considerations with respect to pollinator conservation: foraging plants, very specific nesting habitats such as soft banks for ground-nesting bees, and the fact that bees, for instance, of medium body size can regularly fly up to 2 km between nest sites and foraging patches.

Initiative of Beekeepers

In Nagaland, Mima village has protected about 7 km^2 area as a bee pasture which is located about 600 m away from sub state highway and there are no mankind interventions except well placed honey bee boxes. This area contains a lot of biodiversity for insect pollinators as well as plants also. There, people may not hear other sounds, although one may listen to the melodious buzzing of bees. This initiative shows the love, affection and devotion of Naga tribes with nature, pollinators and honey bees.

Government Policy

Inadequate policies for such an important pollination services exemplify the lack of policy-makers' awareness, and have rarely been addressed in explicit policies to conserve and manage pollination services. The sustainability of ecosystems depends, to a large extent, on the buffering capacity provided by having a rich and healthy diversity of genes, species and habitats. Every country that is party to the Convention on Biological Diversity had committed themselves to develop a National Biodiversity Strategy and Action Plan (NBSAP) because reduction in biodiversity affects these ecosystem services. India specifically refers to pollinators in their NBSAP, in more than one context. Some Indian states and their governments have recognized the role of pollinators; for example, the Himachal Pradesh Department of Horticulture has established departmental bee-keeping stations which are maintained solely for pollination purposes.

Brazil has explicitly promulgated initial policies and intergovernmental directives to support pollination conservation and sustainable use. South Africa's National Biodiversity strategy and action plan recognizes the dependence of production sectors such as cultivation and plantation forestry on ecosystem services, including pollination. In the UK, the National Biodiversity Action Plan1 includes three types of specific action plans; for species, habitats and local planning. Pollination trends and news have been featured in the United States biodiversity information management portal.

Biodiversity Regulations

After adoption of Convention on Biological Diversity (CBD), many countries are in the process of mainstreaming their commitments into national-level biodiversity regulations. While many of these are in draft form, they offer some strong tools for putting pollinator conservation into policy. For example, in Kenya, where ever the biological diversity is not in a protected area, the country's Environmental Management and Coordination Act provides for the Minister of Environment and Natural Resources "to declare any area of land to be a protected natural environment for the purpose of promoting and preserving specific ecological processes, natural environment systems, natural beauty or the preservation of biological diversity in general". Through such measures, a community of coffee farmers, for example, could ask for the protection of a small forest or riparian zone that provides alternative forage and nesting sites to coffee pollinators.

Red Lists

Red Lists are national lists of those insects which are threatened and their population is too low or about to become extinct. For threatened pollinators, they can be very effective tools for guiding policies and local activities to conserve the extinct species. In Europe, there are red lists including bees from Spain, Switzerland, Germany, Great Britain, Lithuania, Latvia, Netherlands, Norway, Poland, Finland and Slovenia. The red list for North America includes 58 bees and 59 butterflies and moths. Many other countries include hoverflies and butterflies in their national red lists.

National Level Initiatives for Pollinator

Brazil has formulated an Understanding for Technical Cooperation between its Environment and Agriculture ministry regarding research on biodiversity and forests, including pollinator conservation and management. Brazilian Pollinator Initiative (BPI) has a unique governmental structure and has been active on many fronts. A national committee of the Brazilian Pollinators Initiative is charged with a number of tasks, *viz.* they have undertaken an inventory of pollination demands of each region of the country for crops with pollination management needs, an exercise that will guide the initiative to focus on priority crops.

Agriculture

In the agricultural sector, pollination has been overlooked in rural development strategies and is not included as a technological input in most agricultural development packages. High value agriculture is promoted by many governments and agricultural development institutions offer packages of practices for different types of crops, but most are overlooking the importance of pollination management to achieve optimized yield. Introducing substantive changes in agricultural development will first require changes in agricultural research and development policies, such as in the research agenda pollination as an important aspect of crop productivity should be recognized and focus on proliferation of their population by augmentation and conservation.

Pollination and Regulation of Agricultural Chemicals

In many countries, there have been efforts to protect honey bees from poisoning by agricultural chemicals but, exposure to the pesticide toxicity on other wild pollinators is rarely considered in agricultural regulations or included in label warnings. About 30 years ago, Rachel Carson wrote the book "Silent Spring", outlining the detrimental effects of pesticides on the environment. Carson warned the world not just about "Silent Springs" but also about "fruitless falls" - in which there is no pollination and subsequently no fruit, due to pesticide poisoning of pollinators.

Habitat Loss

The International Centre for Integrated Mountain Development (ICIMOD) has carried out research on pollination and associated productivity of mountain crops over the past decade. The project has identified loss of habitat and the associated decrease in food and nesting sites for pollinators, resulting from the expansion of agro-ecosystem by replacing the forests-ecosystem and grassland areas, as a major cause to threatening the pollinators and thereby decreased mountain crop productivity. As a result of their findings, the project is making efforts towards conservation of pollinators through raising awareness among farmers and policy makers (Partap and Partap, 2002). The stingless bee population is also decreasing and facing extinction from some areas, and therefore, appropriate measures to conserve their habitat and foraging source as well as natural ecosystem should be emphasized accordingly. These measures include reducing firewood gathering and agricultural expansion, which destroy the ecosystem needed for ground-nesting bees and their floral host plants.

Quarantine Measures

Loss of pollinators can also result from the spread of disease, or through invasive alien species. New Zealand has strict quarantine measures to prevent the introduction of undesirable organisms. The movement of hive bees between North and South Island is strictly prohibited to halt the spread of *Varroa* mites. Although the honey bee is exotic to New Zealand, it is an important agricultural pollinator.

Exotic Bee

The importation of exotic species should be a last alternative. First, try to manage indigenous species and never import species that are known to become invasive. Before the import of any exotic species, ensure that appropriate risk analysis and cost/benefit studies should be undertaken, as in the case of oil palm pollinators (Martins *et al.,* 2003). Recent concern about invasive alien pollinators has focused on planned introductions of *Bombus terrestris*, a European bumblebee widely used for pollinating greenhouse solanaceous crops, such as Tomato, Brinjal and Chilli. Colonies have been exported to Japan, Israel and Chile (*Bombus ruderatus*), and have subsequently naturalised in these regions. In Israel, feral *Bombus terrestris* colonies are a significant ecological threat and also threat to populations of several native bees, including native *Apis mellifera*, showing significant declines (Dafni, 1998).

Rehabilitation of Landscapes

Loss of habitat through land use changes, e.g. due to conversion of natural areas to agriculture, mining or urban development, has been identified as the principal cause of pollinator decline. Farmers can be encouraged to restore some of their farmland to forest or grasslands; road planners can ensure that roadsides and infrastructure servitudes are reseeded with pollinator-friendly plant species, and urban planners can be encouraged to consider native floral diversity in parks.

Apple Pollination in India

In Himachal Pradesh, apple productivity declined continuously for several years due to inadequate pollination. Farmers now use honey bees (*Apis mellifera* or *Apis cerana*) for fulfilment of pollination deficit and provide adequate pollination in apple crop. Some farmers keep their own honey bees while others rent them from the Department of Horticulture or from local beekeepers. At present, only Himachal Pradesh in the entire Hindu Kush-Himalayan region has a well-organised pollination system. This large-scale use of honey bees has led to a new vocation. The success of this enterprise resulted from targeted research into apple pollination by honey bees.

Managing Indigenous Pollinators

Australian government agencies have a long history of investigating the use of honey bees for pollinating crops. Recently, the use of native stingless bees (*Trigona* spp.) for macadamia and cucurbits, and the blue-banded bee, *Amegilla* spp., for pollinating tomatoes in glasshouses, has been instigated. Australia has been able to overcome the need to introduce exotic bumblebees by their native pollinators. Passion fruit (*Passiflora edulis*) growers, especially on smaller farms in Ceara, Brazil, hand-pollinate their crops because its flower is large in size and for their pollination; the only efficient pollinator is carpenter bee (*Xylocopa* spp.) which is insufficient in commercial orchards. It is due to this reason that many farmers have discontinued passion fruit production. Freitas and Oliveira-Filho, 2001 have developed efficient nesting boxes for large carpenter bees and this has increased yield by 92.3 per cent and made hand-pollination unnecessary. Although alfalfa is not native to North America, it is pollinated by a wide array of bees, especially solitary leaf-cutting bees (*Megachile* spp.), of which the alfalfa leafcutting bee is a non-native cultivated species. Many leaf-cutter bees make their nests in tunnels left by wood-boring insects. As a result, nesting habitat for native pollinators diminished and alfalfa growing in large fields remained without pollination and overall seed yields per acre declined. Today, the problem of alfalfa seed production is largely solved by management of "megachileculture".

References

Buchmann, S.E. and Nabhan, G.P. 1996. The forgotten pollinators. Island Press, Washington, DC.

Dafni, A. 1998. The threat of Bombus terrestris spread. Bee World, 79:113-114.

Daily, G. C., Alexander, S., Ehrlich, P.R., Goulder, L., Lubchenco, J., Matson, P.A., Mooney, H.A., Postel, S., Schneider, S.H., Tilman, D. and Woodwell, G.M. 1997. Ecosystem services: benefits supplied to human societies by natural ecosystems. Issues in Ecology: 1-16.

Khan, K.A., Ahmad, K.J., Razzaq, A., Shafiqe, M., Abbasi, K.H., Saleem, M. and Ullah, M.A. (2012). Pollination Effect of Honey Bees, Apis mellifera L. (Hymenoptera: Apidae) on Apple Fruit Development and its Weight. Persian Gulf Crop Protection, 1(2): 1-5.

Martins, D., Gemmill, B. and Eardley, C. 2003. Plan of action of the African pollinator initiative. African Pollinator Initiative, Nairobi.

Partap, U. and Partap, T. 2002. Warning signals from the apple valleys of the HKH: Productivity concerns and pollination problems. ICIMOD, Kathmandu, 124 p.

Pimentel, D., Stachow, U., Takacs, D.A., Brubaker, H.W., Dumas, A.R., Meaney, J.J., O'Neil, J., Onsi, D.E. and Corzilius, D.B. 1992. Conserving biological diversity in agricultural/ forestry systems. BioScience, 42:354-362.

6

Insect Pollinators: Food Security and Biodiversity

Agricultural biodiversity is known as crop genetic resources, although agro-ecosystems hold a wide diversity of other organisms that contribute toward their productivity and sustainability. Amongst these pollinators play the massive role in crop pollination and thus ensure that fruit or seeds are formed. Over 3 decades ago, the international community began to emphasize on the importance of pollinators. In agro-ecosystems, pollinators are essential for agriculture, horticultural and forage production as well as the production of seed for many root and fiber crops. Pollinators such as bees, birds and bats affect 35 per cent of the world's crop production, increasing outputs of 87 of the leading food crops worldwide, plus many plant-derived medicines in our pharmacies (Klein *et al.,* 2007). Food security, food diversity, human nutrition and plant diversity all rely strongly on animal pollinators. Pollination is an essential ecosystem service that enables plant reproduction and food production. The pollinators' abundance and diversity play pivotal role for enhancement of the productivity, quality and maintenance of natural resources of plant biodiversity. The diminution of pollination abundance and their diversity certainly affect productivity, quality and plant biodiversity.

Diversity among angiosperm, including agricultural and horticultural crops, depends on animal pollination are affected with declining pollinators. Pollinator diversity is directly dependent on plant diversity and vice-versa - no other natural phenomenon illustrates the principle more vividly that conservation measures must be directed at ecological processes and not just individual species. Crop-associated biodiversity (CAB) refers to biodiversity that supports the functioning of ecosystem services necessary for agriculture, as well as contributing to the maintenance of ecosystem, health and resilience. CAB is an intrinsic and important part of agricultural ecosystems, and includes components such as pollinators. Pollinators play pivotal role to the maintenance of natural resources of plant biodiversity by virtue of pollination ensued seed production as source of progeny and ensure the survival of plant species including plants that provide food security.

Pollinators

Pollinators are essential for many fruit and vegetable crops pollination, further fruit/ seed set and thereby, enhance the productivity and quality of produce. Food security law can not ensure and enact by making a law, it can ensure by only enhance the crop productivity by maximize pollination. Pollination is an input of ecosystem service for food security. In agro-ecosystem, cross pollinated crops, pollen- scanty crops and heavy pollen bearing crops requires true pollinator for pollination services to enhance the crop productivity, quality and biodiversity. About 80 per cent of plant species pollination requires the help of pollinators; insects, birds (humming bird, bats etc.) and reptiles for transfer of pollen from one flower to another flower of the same species. The principle pollinators are bees, about 73 per cent of the world's cultivated crops have been pollinated by bees, 19 per cent by flies, 6.5 per cent by bats, 5 per cent by wasps, 5 per cent by beetles, 4 per cent by birds, and 4 per cent by butterflies and moths (Freitas *et al.,* 2004).

Although more than 750,000 insect pollinators have been described (Grimaldi and Engel, 2005), possibly as many as 30 million more await discovery and formal description (Erwin, 2004). Approximately 16,000 species of bees are known in the world (Michener, 2000); out of that only a few species have been managed specifically as crop pollinators. The major pollinators belong to Hymenoptera order and out of this mostly belong to Apidae family, other pollinators, such as bats, birds, butterflies, moths, flies and beetles also play key roles in crop pollination. Honey bees (*Apis* spp.) are the most important pollinators of agricultural and horticultural crops. Their morphology and behaviours make them effectual and paramount pollinator for wide a range of angiosperm. Our extent reliability on a single pollinator for such a precious service is much risky. The wild bees are also precious pollinators and our underestimation of their pollination contribution is a blunder. Different crops have different types of flower and according to their morphology, flowers too require specialized pollinators.

Pollinators and Food Security

There are three methods for improving crop productivity; the first method is suitable and sufficient agro-inputs *viz.* good quality seeds, cultural management, manure and fertilizers, irrigation and plant protection. The second method is biotechnological techniques *viz.* manipulating the rate of photosynthesis and exploitation of nutrients result to get high yielding varieties, etc. These biotechnological techniques ensure healthy growth of crop plants but each input is effective up to a limit only. The third is most important but relatively less known method of enhancing crop productivity is to optimize the crop pollination by pollinators (Partap and Partap, 1997). In agriculture, especially amongst cross pollinated, pollen-scanty crops and heavy pollen bearing crops essentially require true pollinators for pollination service.

The pollination process involves the transfer of pollen from anthers to stigma of the same flower is known as self-pollination. The transfer of pollen from anther of one plant to stigma of another plant of the same species is known as cross-pollination. Pollinator is the biotic agent that transfers pollen from anthers to stigma and accomplish fertilization, which is a key ecosystem service for food security. This ecological process is an essential prerequisite for fertilization to result into fruit/seed set. Pollination is therefore the most crucial process in the life cycle of the plants and is essential for crop production and biodiversity conservation and helps enhance farm income and rural livelihoods. Many crops are self-sterile and require pollinators for cross-pollination to produce seeds and fruit (McGregor, 1976; Free, 1993). It is not only self-sterile varieties that benefit from cross-pollination, but rather, self-fertile varieties also produce more and better quality seeds and fruits if they are cross-pollinated (Free, 1993).

The apparent link between pollinators and global food security and biodiversity, declining of managed and wild pollinators are of increasing concern. Food security of increasing population has increasing food requirement and therefore, each and every input play pivotal role towards productivity enhancement. These objectives may be achieved by proliferation of pollinators in ecosystem by augmentation and conservation of pollinators. This concept may fulfill the pollinator deficits in ecosystem and increase the crop productivity fulfillment of proliferated population food security.

Indian government has passed food security bill in parliament and in 2014 the population of our country is about 140 crores. India's population have been proliferating day by day and since food requirement have positive correlation with population, so also, food needs to increase day by day for sustenance. Being a developing country, cultivated land has been encroached gradually for construction of new roads, industrialization, new colonization etc. and it will also continue into the future. For proliferating population, food security should be ensured which ultimately require more food and it may be fulfilled only by enhancement of the productivity because land availability is limited. Augmentation and conservation of pollinators is also best methods to enhance the productivity which may be expected to be a tool of 2nd green revolution as fifth input after land, labor, cost and capital.

Pollinators and Biodiversity

Agro-ecosystem biodiversity is often understood as crop genetic resources, yet it encompasses wide bio-diversity of other organisms that contribute toward their productivity and sustainability. The progress and prosperity of any nation depends upon five factors i.e., land, water, natural resources, technologies, ratio of inflow and out flow currency, environment, forest and biodiversity. Indian occupies the 8th place in biodiversity and is recognized as a 'Biodiversity Hot Spot'. Many ecosystems, including agro-ecosystem and some extended forest ecosystem diversity depend upon pollinator diversity and their abundance. Pollinators belong to diverse groups of the animalia kingdom, including birds, bats, reptiles, insects, etc. Conservation of pollinators' biodiversity is important for the potential yield of agricultural as well as horticultural crops and their hybrid seed production. Pollinators play an important role in conservation of plant biodiversity as well as survival of several plant species through mixing of gene pool. India is endowed with the greatest biodiversity of insect pollinators.

In recent years the Convention on Biological Diversity (CBD) has recognized pollination as a key driver in the maintenance of biodiversity and ecosystem function. More diverse communities of pollinators in agricultural systems also have greater total abundances and rates of visitation to crop flowers (Pritchard, 2005; Chacoff and Aizen, 2006). Therefore, maintaining diverse pollinator communities locally is important for providing pollination services to a more diverse set of crops. Within a crop, a diverse group of pollinator species can provide better pollination services than a single species. The 25,000 different bee species, size and habit significantly differ from each other and differ accordingly in the plants they visit and pollinate; however, the co-evolution of pollinators also depends upon the related fauna. The diversity of fauna and their variability of food source also depend on the co-evolution of pollinators and their diversity. The principles and best practices described below are all aimed at addressing these particular needs, and can be recommended to conserve the biodiversity.

How to Overcome

- Plant-pollinator relationships should be understood as an ecosystem service for sustainable agriculture. This includes a concerted plan to overcome the taxonomic impediment.
- Augmentation and conservation of pollinators.
- Conservation of natural ecosystem/ buffer zone as a congenial niche for pollinators to conserve pollinators and optimize pollinator services

in agro ecosystems. Particular attention should be paid to protection of appropriate nesting sites.

- Judicious use of agro-chemicals; if it is necessary, select bio-rational and selective chemicals but application should be avoided in blooming period and it may be applied at evening time when pollinator activity ceases i.e. cessation of foraging.
- Farming practices; minimum admissible tillage of soils will assist to conserve the nests of pollinator.
- Pollinators congenial farming practices should be promoted to conserve its abundance and diversity, through the following envisage:

A On the one hand there is small-scale agriculture amidst undisturbed, natural areas from which pollinators may migrate onto the agricultural plots and which should be conserved and protected with concern.

B Avoid mono-cropping system and incorporate those crops which may bloom during dearth period.

C Public awareness about the importance of pollinators and their conservation practices should be promoted.

Declining Pollinators and Increasing Population

Undoubtedly agricultural and horticultural crops production, productivity and diversity are diminishing with declining populations of pollinators and their extinction. Many pollinators' population have been decline and the available population are not able to fulfil the required pollination. Therefore, crop productivity and diversity of dependent crops are in peril. The declining trend of pollinators' abundance and diversity are of major concern for ecosystem service to sustain the crop productivity and plant biodiversity. The ecosystem is suffering from scarcity of pollination due to modern agricultural practices, habitat fragmentation which accompany decrease in their food (nectar and pollen), monoculture which elongate the floral dearth period, use of pesticides which expose pollinators to toxicity, erosion of their nesting habitats which destructs the niche and climate change which affect pollinators and their desire *viz.* food, habitat and congenial weather.

In 2023, our country's population is about 140 crores and the population is increasing day by day where food requirement have positive correlation with population and therefore, food production too needs to be increased daily for sustenance. Being a developing country, cultivated land has continuously been encroached for construction of new roads, industrialization, new colonization, etc. and it will also continue in future. For the proliferating population, food security should be ensured which ultimately require more food and it may be fulfilled only by enhancement of the productivity because land availability is limited. Augmentation of pollinators and sufficient pollination is also one of the best methods to enhance the productivity. This increasing pressure on supply of agricultural land could significantly contribute to global environmental change.

Impact of Declining Pollinators and Their Diversity on Oroductivity

The decline in pollinator population and diversity are a major concern to agricultural productivity and their sustainability. There is clear cut indicator on pollinator dependence that plant productivity has been decreasing in those areas where pollinators' population and diversity are declining despite necessary agronomic inputs. Plant productivity has been declining due to deficit of pollination. Pollination deficits occurs because all the ovules may not fertilize and ensued low number of fruit and seed form and do not obtain potential yield. Rented pollinators colony placed in the flowering fields for optimizing the pollination is also a signal of pollinator scarcity, e.g., Himachal Pradesh and Jammu farmers are complaining of declining apple productivity despite providing the best agricultural inputs and the same trend is happening in northwest India, northern Pakistan and parts of China in almonds, cherries and pears crops. For fulfillment of pollination deficit, farmers have been trying to augment pollinator deficit by placing *Apis mellifera* colony in apple orchards on rental basis. As a result of low yields and quality of apples from poor pollination, several farmers have chopped off their apple trees (Partap, 2001).

Implication of the decline in the pollinator populations as well as diversity is that it has created the need for managed pollination in order to maintain crop yields and quality. In Hengduan Mountains of China, farmers have been pollinating their crops, e.g. apples and pears through hand pollination using human beings as pollinators (beekeepers do not rent their honey bee colonies for pollination of these crops because farmers make excessive use of pesticides even during flowering season). Hand pollination is an expensive, time-consuming and highly unsustainable proposition of crop pollination owing to the increased labour scarcity and costs.

How to Ensure Food Security

Floral ecology and pollinators' conservation are new concerns within the ecosystem which has not been explored adequately. On World Food Day

2004, the FAO stressed that safeguarding and using the potential and diversity of nature is critical for world food security. The role of pollinators towards achieving food security is one critical function to all humanity and full attention to it is long overdue. The world population is growing rapidly and the demand of food production will be double in the next thirty-five years. The cultivated land is limited and it has been continuously encroached by new constructions and therefore, the only one way to overcome this challenge is to increase the productivity. This raises serious concerns about food security - how will we produce enough food to fulfill our required food demand in the future? The major possibilities of agricultural inputs have been applied *viz.* high yielding new varieties, effective water management, fertilizers and plant protection methods, but there are limits of these inputs up to some extent to increase crop productivity. An additional vital ingredient underpinning our ability to enhance the productivity with maximize the pollination. It is only by proper management of pollination and pollinators that would be able to enhance our food productivity and overcome the forthcoming challenge and provide food security in the coming years. For instance, eight out of ten European crops depend on insect pollination. Most of these pollinators are bees, including managed honey bees and a wide range of wild bees such bumblebees as well as hoverflies. Most of our crops could not produce seeds and fruits without bees pollination.

Another side, our food varieties depend on pollinators. Such as our breakfast: salads, fruit juice, fresh fruits, jam, marmalade, coffee, chocolate and many more rely on pollinators. Most of our herbs and spices are also pollinator dependent and these bring flavours and make our meals yummy. Even meat and dairy products indirectly depend on the pollinators, because some of their foods like mustard cake and fodders Alfa alfa grass and Barsim seed production depends upon pollinators. Therefore, in the absence of pollinators, our diets would become greatly restricted; many fruits and vegetables would be off the menu. Over the past few decades, there has been a significant loss of pollinators including honey bees, native bees, birds, bats, and butterflies from the environment. The problem is serious and poses a significant challenge that needs to be addressed to ensure the sustainability of our food production systems and avoid additional economic impacts on the agricultural sector. Insects are an essential part of ecosystems and insect pollination and so, maintaining pollinator biodiversity is essential for food security. Globally, 87 of the leading 115 food crops evaluated are dependent on animal pollinators, contributing 35 per cent of global food production.

Pollinators' pollination is a vital ingredient in food, providing choice, healthy diets and security. By protecting key components of biodiversity we ensure that we can meet some of the most basic needs of society. In Australia, Wheen Bee Foundation mission is to improve national food security by striving to ensure a viable and prosperous beekeeping industry. The vision and mission of this foundation is to ensure healthy honey bees and a prosperous beekeeping industry that will provide pollination services for the production of high quality and nutritious food to achieve global stability through food security.

How to Conserve the Biodiversity

The International Initiative for the Conservation and Sustainable Use of Pollinators was established by the 5th Conference of Parties to the Convention on Biological Diversity (CBD) as a cross-cutting initiative within its work on agricultural biodiversity. FAO facilitated and co-ordinated the initiative in co-operation with other relevant organizations who were invited to support actions in countries subjected to pollinator decline. The FAO is Global Action on Pollination Services for Sustainable Agriculture provides guidance to member countries and relevant tools to use and conserve pollination services that sustain agro-ecosystem functions. Scientists have identified about 1.4 million unique species of plants and animals on the planet and about one new species is added to this list every day. We depend on pollinators for our food, energy, shelter and in numerous other ways as well. But as the planet's human population proliferates, the biodiversity is coming under increasing threat.

Pollination is essential for the maintenance of diversity in wild flowers and is indirectly responsible for the persistence of other guilds that depend upon floral resources such as herbivores. However, the economic, biodiversity and aesthetic value of pollinators is known for relatively few systems (Nabhan and Buchmann, 1997; Delaplane and Mayer, 2000), and there is considerable scope to improve our understanding of the habitat characteristics that moderate the value of plant-pollinator communities. The biodiversity value of these habitats for pollinators has received some attention (Petanidou and Ellis, 1993; Potts *et al.,* 2003), but the services that pollinators provide to flowering plants is still relatively unexplored despite the fact that many wild plants experience significant pollen limitation (Burd, 1994). Researches may provide specific data that strengthen the need for the conservation of different plants and animals, and offer subsidies to propose necessary information for the execution of public and private policies aimed at conservation of the biodiversity.

Considering pollination solely as a service for human food consumption is unwise, because it indirectly affects biodiversity of a decline in pollinators and also, affects human welfare (Kremen *et al.,* 2007). Many people value biodiversity for cultural reasons that extend far beyond the biological processes which depend on it. The potential consequences of pollinator decline leads to loss in conservation of biodiversity and instability of food crop yields (Allen-Wardell *et al.,*1998).

Augmentation in Substitute of Inadequate Pollination

Undoubtedly agricultural and horticultural crop production, productivity and agro-ecosystem diversity are threatened by declining populations of pollinators and their extinction. Many pollinator population densities have been reduced and the available population are unable to fulfil the required pollination service of plants and therefore, plant productivity and diversity of dependent crops are in peril. In agro-ecosystem, many crops depend upon insects for their pollination, and assisted pollination may have to be done when self pollination is insufficient in order to reduce potential yield loss (Klein *et al.,* 2007). Inadequate pollination is a major constraint to crop yield. Therefore, pollination augmentation or managed crop pollination is a core input and solution in consideration of deficit pollination to obtain optimum crop productivity.

Nowadays some innovative farmers are using rented insect pollinators in their orchard for fulfilment of pollinator deficits to obtain the productivity in the valuable cross pollinated crops, which depend upon insect pollinators. Augmentation of insect pollinators is an effort to increase population of insect pollinators either by proliferation or by environmental manipulation. Augmentation is the placing of additional population of naturally occurring pollinators into the ecosystem to fulfill the deficits of pollination.

Pollination is an important service that improves the quantity and quality of many agricultural crops such as apples, almonds, melons, blueberries, strawberries and alfalfa. Managed pollination of crops has been a neglected part of agriculture, which requires due attention to increase productivity and quality. In USA, Canada, Germany etc., millions of honey bee colonies are hired on rental basis for pollination services in valuable commercial crops. In India, it is also adopted in apple orchards in Himachal and Jammu Kashmir. For some crops, e.g., apples, almonds, and blueberries, pollination has long been a managed activity with bee colonies being transported from location to location to coincide with the flowering season.

Challenges of Pollinator Augmentation

Although more than 750,000 insect pollinators have been described (Grimaldi and Engel, 2005), possibly as many as 30 million more await discovery and formal description (Erwin, 2004). Approximately 16,000 species of bees are known in the world (Michener, 2000) and only a few species, however, are managed or "kept" specifically as crop pollinators' *viz.* honey bees, stingless bees, bumblebees and solitary bees. These pollinators have been used in North America, Europe, Japan, China and India and these pollinators have been in use to ensure pollination in cross pollinated and valuable horticultural crops. Although rearing and using species of bumblebees, stingless bee and solitary bees are practiced, there is no standard practice of using hives like bees such as *Apis cerana* and *Apis mellifera.* The stingless bees are reared in traditional bee hives throughout the world. Australia has evaluated some designed of stingless bee boxes. In India, AICRP (HB&P) Nagaland, AICRP (HB&P) Kerala have designed stingless bee boxes. These boxes show great promising results. Stingless bees are very good pollinators under net house & Greenhouse condition due to low flight range. Only honey bees are reared and used as a rented pollinator in scientific way while all the other pollinators are still unexplored due to lack of knowledge. As a whole, to integrate pollination with farming systems and enhancing rural livelihoods by promoting managed pollination and conserving pollinator populations is also a major challenge. In India, only limited people and farmers know about the importance pollinators. The main constraints are lack of awareness and understanding among farmers, extension workers, planners and policy-makers about the importance of pollinators and pollination.

References

Allen-Wardell, G., Bernhardt, P., Bitner, R., Burquez, A., Buchmann, S., Cane, J., Cox, P. A., Dalton, V., Feinsinger, P., Ingram, M., Inouye, D., Jones, C. E., Kennedy, K., Kevan, P., Koopowitz, H., Medelllin, R., Medellin-Morales, S., Nabhan, G. P., Pavlik, B., Tepedino, V., Torchio, P. and Walker, S. 1998. The potential consequences of pollinator declines on the conservation of biodiversity and stability of food crop yields. Conservation Biology, 12(1):8-17.

Burd, M. 1994. Bateman's principle and plant reproduction: the role of pollen limitation in fruit and seed set. Botanical Review, 60: 83-139.

Chacoff, N. and Aizen, M. 2006. Edge effects on flower-visiting insects in grapefruit plantations bordering premontane subtropical forest. Journal of Applied Ecology, 43: 18-27.

Delaplane, K. S., Mayer, D. F., 2000. Crop Pollination by Bees. CABI Publishing, UK.

Erwin, T. L. 2004. The biodiversity question: How many species of terrestrial arthropods. Forest Canopies, 2nd edition, M.D.L. Lowman, and H.B. Rinker, eds. Burlington, Vt.: Elsevier Academic Press 259- 269.

Free, J. B. 1993. Insect pollination of crops, 2nd edn. Academic, London.

Grimaldi, D. and M. S. Engel. 2005. Evolution of the Insects. New York: Cambridge University Press.

Klein, A-M., Vaissi-Are, B. E., Cane, J. H., Steffan-Dewenter, I., Cunningham, S. A, Kremen, C. and Tscharntke, T. 2007. Importance of pollinators in changing landscapes for world crops. Proceedings of the Royal Society B: Biological Sciences 274: 303-313.

Kremen, C., Williams, N. M., Aizen, M. A., Gemmill-Herren, B., LeBuhn, G., Minckley, R., Packer, L., Potts, S. G., Roulston, T., Steffan-Dewenter, I., Vazquez, D. P., Winfree, R., Adams, L., Crone, E. E., Greenleaf, S. S., Keitt, T. H., Klein, A-M, Regetz, J. and Ricketts, T. H. 2007. Pollination and other ecosystem services produced by mobile organisms: a conceptual framework for the effects of land-use change. Ecology Letter, 10:299-314.

McGregor, S. E. 1976. Insect pollination of cultivated crop plants. United States Department of Agriculture (USDA), Washington DC.

Michener, C. D. 2000. The bees of the world. The John Hopkins University Press, Baltimore/ London, 913 pp.

Nabhan, G. P. and Buchmann, S. 1997. Services provided by pollinators. In G. C. Daily (Ed.), Nature's services: Societal dependence on natural ecosystems (133-150). Washington, DC: Island Press.

Partap, U. 2001. Warning signals from the Apple Valleys. Video Film, 31 minutes and 7 seconds, VHS Format, ICIMOD, Kathmandu.

Partap, U. and Partap, T. 1997. Managed crop pollination: the missing dimension of mountain agricultural productivity. Mountain Farming Systems' Discussion Paper Series No. MFS 97/1, ICIMOD, Kathmandu.

Petanidou, T., Ellis, W.N., 1993. Pollinating fauna of a phryganic ecosystem: composition and diversity. Biodiversity Letters, 1, 9-22.

Potts, S. G., Vulliamy, B., Dafni, A., Ne'eman, G., O-Toole, C., Roberts, S., Willmer, P. G., 2003. Response of plant-pollinator communities following fire: changes in diversity, abundance and reward structure. Oikos, 101, 103-112.

Pritchard, K. 2005. The unseen costs of agricultural expansion across a rainforest landscape: Depauperate pollinator communities and reduced yield in isolated crops. Unpublished master's thesis, James Cook University, Queensland, Australia.

7

Pesticides and Pollinators

Pollinators are organisms that help to transfer pollen from anther to stigma of the same species for fertilization and thereby produce seeds and fruits. Pollinators belong to diverse groups viz. insects, birds, reptiles, mammals, etc. Insects; bees butterflies, moths, wasps, flies, some types of beetles. Out of insect pollinators, bees are most important pollinator. Insect pollinators are critical to the production of many crops and play pivotal role to maintain the biodiversity. The decline in honeybee and other pollinator populations have more attention of scientific and public concern globally as well as in India. A number of factors have been observed as potential contributors to these declines and no single factor has been identified as the cause. The available science suggests that multiple factors responsible for population depletion and extinction, including loss of habitat and food sources, diseases, viruses, pests, and pesticide exposure. It is also known that certain pesticides can pose an immediate or "acute" threat to the insect pollinators and bees. In order to protect pollinators, the labels of pesticides that pose such risks should specify detailed directions of application to reduce potential exposure.

Plant protection products (pesticides) contain biologically active compounds used against insect pest in order to plant protection. Insecticides are used to control insect pest populations; herbicides control weeds while fungicides are used to control fungal plant diseases. Pesticides are used in agriculture, horticulture, and forest ecosystem to control a wide range of insect pests, acarids, micro-organism and weeds. About 90 registered pesticides have been identified as toxic to bees in India. The application of these pesticides and their exposure may come into contact with pest and insect pollinators as well as bees. Unfortunately, honey bees are insects and are greatly affected by insecticides. Those pesticides which are applied during flowering stage to control the crop pest has been revealed to be more hazardous for insect pollinators because simultaneously pollinators visit for foraging and pollinating the flowers and are exposed to pesticides in the treated area. Out of the pesticides, insecticides are highly hazardous for insect pollinators. Among the insecticide groups, organophosphates, carbamates, synthetic pyrathroids and neonicotinoids are more hazardous for insect pollinators. Out of these, neonicotinoids group is highly hazardous for pollinators in comparison to other groups due to its systemic and long persistent nature.

Neonicotinoids embarked in Indian market during the first decade of this century whereas it was available in the developed countries since the mid 1990s. Neonicotinoids are synthetic chemical and systemic insecticides with their structure and mode of action similar to Nicotine. European states have voted in favour of a proposal to restrict the use of pesticides linked to serious injury in bees. Europe is set to enforce the world's first continent-wide ban on widely used insecticides alleged to cause severe injury to bees, after the European commission votes on the matter. The hung vote hands the final decision to the European commission which will implement the ban. The commission proposed the suspension after the EFSA concluded in January that three neonicotinoids; Thiamethoxam, Clothianidin and Imidacloprid posed an unacceptable risk to bees. The three will be banned from use for two years on flowering crops such as corn, oilseed rapeseed and sunflowers, upon which bees forage. In USA, Freedom of Information Act has called for a ban on neonicotinoid use for seed treatments because of their toxicity to birds, aquatic invertebrates and other wildlife (Mineau and Palmer, 2013).

Augmentation of insect pollinator to fulfill the pollination deficits and maximization of productivity practices are required on crop in particular areas for pollination purposes. On the other side, simultaneous application of pesticides is a need as a tool of IPM to minimize crop loss due to pest. Both practices are essential to optimize the crop productivity and therefore, use of conscience is required to amicably practice both processes in the same field with slight modification, *viz.* determination of pest ETL, selection of pest management tools from IPM modules, selection of pesticides, selection of pesticide formulation, selection of pesticide application time etc. to minimize the hazardous effects on pollinators. There are several ways through which honey bees are killed by insecticides. One is through direct contact of the insecticide to the bee while it is foraging in the field. The bees immediately die and do not return to the hive. In this case the queen, brood and nurse bees are not contaminated and the colony survives. In the second way, bees come under contact on treated area during foraging, collect the toxicated nectar and pollen and store them in bee hive which is very harmful for bee colony and may even lead to collapse of the colony. Therefore, before application of pesticide for crop protection we must think and consider that there are so many economic and efficacious tool available for pest management as option but pollinators have only one option for flower foraging.

All precautions and directions on pesticide labels should be followed meticulously. The government should also ensure that agricultural practices across the country protect pollinators by enacting special regulations. It should also collaborate with other pesticide regulators internationally to refine pesticide risk assessment methods and data requirements so that the potential effects on honey bees and pollinators are better understood and risks can be mitigated.

Pollinators and Herbicides

Insect pollinators completely depend upon angiosperms for their food and to some extent, for nest. They positively influence the plants pollination and plants provide food source and shelter (habitat) while some of these plants may also be weeds. Pollinators also depend upon weeds for food source but uses of herbicide exterminate weeds. Applications of herbicides destroy the food source of pollinators and its application at flowering stage generate toxic food source for a limited time which is hazardous for pollinators. MSMA, paraquat, and cacodylic acid are also highly toxic when sprayed onto older bees in small cages (Moffett *et al.,* 1972).

Pollinators and Fungicides-

Fungicides alter the foraging behavior of insect pollinators and can be toxic to adults or larvae. The bees also stop foraging on flowers sprayed with fungicides. The bees, according to studies, collected pollen sprayed with fungicide during their pollination activities. The propiconazole is toxic and reduce the survivorship of adult honey bee. One study found that propiconazole mixed with a pyrethroid insecticide was 16.2 times more toxic to honey bees than the insecticide alone. Exposure to pollen containing captan, ziram led to 100 per cent mortality of larvae. Most fungicides are directly harmful to bees or indirectly affect bees' food source by killing gut microflora and cause gastroenterology disorders.

Pollinators and Insecticides

Insecticides have been divided into different groups and each group of insecticide has different hazardous effects on insect pollinators. The different group of insecticide and their effects on pollinators are given below:

Pollinators and Organophosphates

These pesticides affect the nervous system by disrupting the enzyme that regulates acetylcholine, a neurotransmitter leading to overstimulation and dysfunction of the nervous system. The organophosphates active ingredients are classified (many ending in -phos, -fos, -vos, or -thion). These compounds have a wide range of toxicity levels and are chemically similar to nerve gases developed for human warfare. Damage to brood and queen by exposure to microencapsulated methyl parathion or acephate have also been recorded. The symptoms of organophosphate poisoning in honey bees include loss of activity, abnormal wobbly movements, lying on the back or spinning while beating wings in this position and regurgitation of collected nectar.

Pollinators and Carbamates

Carbamate affects the nervous system by disrupting an enzyme that regulates acetylcholine, a neurotransmitter. The enzyme effects are usually reversible. There are several subgroups within the carbamates; N-methyl carbamates, thiocarbamates, and dithiocarbamates. The names of many of these active ingredients end in the suffix -carb, and the class also includes several insecticides that are responsible for many bee poisonings (carbaryl, carbofuran). Like the organophosphates, they are inhibitors of acetylcholine metabolism in the nervous system and thus share similar symptoms. Symptoms of carbamates poisoning in honey bees include inability to fly in adult bees, dead brood or newly emerged dead workers or queen loss. Sublethal effects on the queen have also been recorded, such as poor or erratic egg laying performance.

Pollinators and Synthetic Pyrathroids

Pyrethroid insecticides are a class of synthetic compounds based on the naturally occuring compound pyrethrin. Pyrethrin is noted for its quick 'knock-down' of insects, but the natural compound is not always lethal, and degrades readily in the environment. Pyrethroids inhibit the nervous system of insects. This occurs at the sodium ion channels in the nerve cell membrane. Some pyrethroids also affect the action of a neurotransmitter called GABA (3). Altering the amount of ions (charged atoms) passing through ion channels causes the membrane to depolarize which, in turn, causes a neurotransmitter to be released. Neurotransmitters help nerve cells in communication. Electrical messages sent between nerve cells allow them to generate a response, like a movement in an animal or insect. Symptoms of pyrethroid poisoning in honey bees include regurgitation of collected nectar.

Pollinators and Neonicotinoids

Neonicotinoids are synthetic chemical. Their structure and mode of action resembles Nicotine and act as neuro-active insecticides. This group of insecticides embarked in Indian market during the 1st decade of this century whereas it was available in the developed countries in the mid 1990s. Their residues are also present in nectar and pollen which are main source of ood for pollinators and therefore, it causes direct threat to bees and other pollinators. The European commission proposed the suspension after the EFSA (European Food Safety Authority) concluded that three neonicotinoids; Thiamethoxam, Clothianidin and Imidacloprid posed an unacceptable risk to bees. The three will be banned from use for two years on flowering crops such as corn, oilseed rape and sunflowers upon which bees forage for food. Honey bees exposed to sub-lethal level of neonicotinoids can affect its flying, navigation and reduce test sensitivity which altogether affects foraging ability. Bumble bees exposed to sub-lethal level of neonicotinoids exhibit reduced food consumption, reproduction, worker survival rate and foraging activity.

Major insecticide in India and their relative toxicity for honey bee-

S.No.	Common name of the insecticide	Class/ group of the insecticide	Level of toxicity
1	Carbaryl	Carbamates	Highly toxic
2	Methomyl	Carbamates	Highly toxic
3	Carbofuron	Carbamates	Highly toxic
4	Carbosulfan	Carbamates	Highly toxic
5	Monocrotophos	Organophosphate	Highly toxic
6	Chlorpyrifos	Organophosphate	Highly toxic
7	Acephate	Organophosphate	Highly toxic
8	Traizophos	Organophosphate	Highly toxic
9	Profenophos	Organophosphate	Highly toxic
10	Dimethoate	Organophosphate	Highly toxic
11	Cypermethrin	Synthetic pyrethroids	Highly toxic
12	Deltamethrin	Synthetic pyrethroids	Highly toxic
13	Bifenthrine	Synthetic pyrethroids	Highly toxic
14	Lambda-cyhalothrin	Synthetic pyrethroids	Highly toxic
15	Pyrethrum	Botanicals	Low toxic
16	Azadirachtin	Botanicals	Low toxic
17	Diflubenzuron	Insect growth regulator	Non toxic
18	Buprofezine	Insect growth regulator	Low toxic
19	Methoxyfenozide	Insect growth regulator	Non toxic

20	Novaluron	Insect growth regulator	Highly toxic
21	Acetamiprid	Neonicotinoids	Low toxic
22	Clothianidin	Neonicotinoids	Highly toxic
23	Imidacloprid	Neonicotinoids	Highly toxic
24	Thiacloprid	Neonicotinoids	Low toxic
25	Thiamethoxam	Neonicotinoids	Highly toxic
26	Dinotefuron	Neonicotinoids	Highly toxic
27	Avermectin	Macrocyclic lactones	Highly toxic
28	Emamectin benzoate	Macrocyclic lactones	Highly toxic
29	Spinosad	Macrocyclic lactones	Highly toxic
30	Flubendiamide	Diamides/ Antraknil	Non toxic
31	Chlorantraniliprole	Diamides/ Antraknil	Non toxic
32	Fipronil	Phenyl pyrazole	Moderatly toxic
33	Indoxacarb	Oxidiagine	Moderatly toxic
34	Metaflumizone	Semicarbozone	Moderatly toxic
35	Cartap Hydrochloride	Cartap Hydrochloride (Nereistoxin)	Low toxic
36	*Bacillus thuringiensis*	Bio/ bacterial group	Low toxic

The untreated plant can also absorb living residue from soil and become a sub lethal for insect but lethal for bees. Bees have genes for specific types of nicotinic acetylcholine receptors (Jones, 2006), and therein may lie the special sensitivity they have to neonicotinoids, but behavioral outcomes of selective actions at these molecular targets has yet to be investigated. Bee kills in France and Germany have been associated with particularly imidacloprid (Halm *et al.,* 2006) and clothianidin (Everts *et al.* 2004). Neonicotinoids and systemic fungicides are often combined as pest control inputs, and many of the latter synergize the already high bee toxicity of neonicotinoids (Iwasa *et al.,* 2004). In 2012, several peer reviewed independent studies were published showing that neonicotinoids had previously undetected routes of exposure affecting bees including through dust, pollen, and nectar (Tapparo *et al.,* 2012).

Pollinators and Fipronil

Fipronil is a phenylpyrazol insecticide is used for pest control in agriculture that mode of action differs from that of the organophosphorus and carbamates (both cholinesterase inhibitors) and some pyrethroids (sodium channels activators) whereas fipronil interferes with the function of aminobutyric acid (GABA)-gated channels. Fipronil is highly toxic to nontarget organisms, such as aquatic species, terrestrial game birds, and honey bees (Le Faouder *et al.*, 2007). Honeybees can encounter by fipronil during foraging by disturbance of different physiological action. Patch-clamp recordings of insect neurons revealed that fipronil blocks both GABA and glutamate-gated-chloride channels (Barbara *et al.*, 2005; Janssen *et al.*, 2007). This treatment decreased acquisition success and the subsequent memory performances were lowered but the distribution of responses to the tactile stimuli between sides was not affected (Bernadou *et al.*, 2009). The *Nosema ceranae* with fipronil combination led to a significant decrease in honeybee survival compared to control or single treatments (Aufauvre *et al.*, 2012).

Major acaricides in India and their relative toxicity for honey bee-

S.No.	Common name of the acaricide	Class/ group of the acaricide	Level of toxicity
1	Properzite	Acaricide/ Sulfite Ester	Moderately toxic
2	Fenazaquin	Acaricide	Moderately toxic
3	Spiromesifen	Acaricide	Moderately toxic
4	Abamectin	Acaricide/Avermectins	Highly toxic

Emamectin Benzoate

Emamectin is widely used in controlling caterpillars of Lepidoptera order pests. It has been shown to possess a greater ability to reduce the colonization success of engraver beetles and associated wood borers. It works as a chloride channel activator by binding gamma aminobutyric acid (GABA) receptor and glutamate-gated chloride channels disrupting nerve signals within arthropods (Grant, 2002). The stronger binding of GABA increases the cells permeability to chloride ions within the cell due to the hypotonic concentration gradient and neurotransmission is thereby reduced by subsequent hyperpolarisation and the elimination of signal transduction. It is highly toxic for honey bees, *Apis mellifera* LD50 (24 h) oral = 0.0036 µg/ bee. It is highly toxic for lepidopterous and coleopteran pollinator. Emamectin benzoate insecticides proved to be comparatively safer than other clorpyriphos and imidachloprid.

Major fungicides in India and their relative toxicity for honey bee-

S. No.	Common name of the fungicide	Class/ group of the fungicide	Level of toxocity
1	Mancozeb	Ethylene bisdithiocarbamate	Low toxic
2	Propineb	Ethylene bisdithiocarbamate	Low toxic
3	Carbendazim	Benzimidazol	Low toxic
4	Tebuconazole	Traizol	Low toxic
5	Pencycuron	Phenylurea	Low toxic
6	Flusilazole	Traizol	Low toxic
7	Copper Hydroxide	Inorganic copper	Moderately toxic
8	Metalaxyl	Acylalanine	Slightly toxic
9	Hexaconazole	Triazole	Low toxic
10	Metiram	EBDC group	Low toxic
11	Dimethomarf	Urea	Low toxic
12	Pyraclostrabin	stroblin	Non toxic
13	Propiconazole	Triazole	Low toxic
14	Chlorothalonil	Chlorinated isophthalic acid derivative	Low toxic
15	Cymoxanil	Cyanoacetamide oxime	Low toxic
16	Metiram + Pyraclostrabin	EBDC group+ Stroblin	Non toxic

Ways of Pesticide Poisoning on Insect Pollinators

- Pesticides poisoning on the insect pollinators generally occurs after application of pesticide at blooming stage on crops or weeds where treated flowers provide toxicated nectar and pollen as a food source to the pollinators.
- The pesticides are exposed directly on the foraging source of pollinators.
- Pollinators make contact with the treated plants and collect toxicated nectar or pollen or both.
- Pollinators collect contaminated water from the near to treated fields.
- During pollen collection, pollinators collect dust formulation of pesticide with contaminated pollen and contaminated nectar with pesticide residue and store in their nest.
- Application of toxicants to control the honey bee pest in hive is also harmful for honey bees.

Field Symptoms of Pesticide Poisoning on Honey bee

- Affected bees may show uncoordinated movement or behave abnormally, abnormal communication, paralysis and eventually death.
- Some pesticides, particularly systemic pesticides, have a less noticeable but debilitating effect, resulting in an overall dwindling of the colony. The reduction in worker adult bee numbers and missing stages of the brood cycle or complete brood cycles are the effects of the pesticides.
- Most commonly, large numbers of forager bees are found dead at the front of bee hives on the bottom board and ground.
- Most or all of the hives may be affected and adult bees' population decrease within a few days.
- Dead adult bees typically, but not always, have their wings unhooked at odd angles to their body, their proboscis fully become outside and their hind legs outstretched behind them. Different group of pesticides give different symptoms.

Recovery from Pesticide Exposed bee hives

Colonies that have been exposed to pesticides may recover if proper steps are taken as mentioned below:

- Bee hives must be transferred to a safe area.
- Removing the super chamber may help to maintain the hive temperature especially in cold condition.
- Provide sugar syrup until the colony becomes sufficient in strength again.
- Problems may occur for a number of days after the pesticide application. Be prepared to manage the hives for queen failure by placing a new queen or through supersedure.
- If brood and nurse bees continue dying without symptoms of diseases and parasites, it shows that the pesticide is still present in the hive, probably in the stored pollen or raw honey or both. The colony will continue to die as long as the poison remains in the hive. In these cases, the combs should be removed and new frame as well as combs should be placed again.

How to Reduce the Bee Poisoning by Pesticides

- Apply pesticides only when it is essential.
- Choose bio-rational pesticides with the lowest hazard rating for bees, particularly selective, high desipation and lowest residual toxic effect.
- Liquid or granule applications are less hazardous than dusts formulation.
- Ground application is less hazardous than aerial application, particularly when applied in close distance in apiaries.
- Application of pesticide must be in the evening after cessation of bee foraging.

Preventive Measures

Precautionary practices may minimize the insect pollinators' toxicity:

Selection of Pesticides

Choose the selective, bio-rational pesticides which are less toxic for insect pollinators and which degrade rapidly. Many of the newer pesticides being marketed today degrades faster and are safer up to some levels for bee activity. When these pesticides are sprayed in the fields, it takes only a few hours for them to degrade as opposed to a few days or weeks.

Selection of Pesticide Formulation

The appropriate choice of formulation is another way to minimize the pesticide exposure to pollinators. Pesticides come in different formulations; solutions, emulsifiable concentrate and granular are the best formulations to minimize the risk of pesticide exposure. Solutions and emulsifiable concentrates dry quickly and do not leave a powdery residue unlike the dusts and wettable powders. Granules are similar to dusts but are larger in particle size and applied into the soil or broadcast on the surface of the ground. On the other hand, dusts and wettable powders will adhere on the pollinators' body together with pollen and are stored in their nest. This cause health hazard for adult as well as brood also.

Time of Application

Many pesticides are extremely toxic to honey bees and other insect pollinators. Honey bees are attracted to flowers for foraging. If at all possible, do not spray pesticides directly during blooming stage but if this stage needs to be sprayed, apply the pesticides in the evening hours after cessation of pollinators foraging. Mostly, pollinators' forage during daylight hours and their foraging have positive correlation with light intensity. As the sun begins to set, they cease their foraging activity. Thus, spraying pesticides in the evening hours can greatly reduce hazards on pollinators.

Alter Application Method

The method of application can also change the risk of pesticide poisoning. Aerial applications have the highest potential risk for exposure on the pollinators. Most pollinators are killed when the pesticides drifts onto non-targeted pollinators' attractive crops. The outcome of such drifts can be catastrophic. Spraying during windy days greatly increases the risk of drift. Using granular formulations, soil treatments or equipment that confines the spray to the intended target can help reduce the risk of drift from pesticides.

References

Moffett, J.O., Morton, H.L. and Macdonald, R.H. 1972. Toxicity of some herbicidal sprays to honey bees. Journal of Economic Entomology, 65:32-36.

Pierre, M. and Cynthia, P. 2013.The Impact of the Nation's Most Widely Used Insecticides on Birds. Neonicotinoid Insecticides and Birds. American Bird Conservancy Publication.

8

Importance of Insect Pollinators in Sustainable Agriculture

In agro-ecosystem, many agricultural and horticultural crops are dependent on insects for their pollination; their degree of interdependence varies species to species. Some plant species depend primarily on a single pollinator's species or genus, whereas some plant species accept diverse pollinators. About to 80 per cent of flowering plants' pollination determined for pollinators (FA0, 2007) and out of that most of the floras depend on insect pollinators for pollination and pollinators depend on flora for their food. Bees are considered as most important pollinator due to their diversity, credential ability and effective pollinators of many plant species. Most of the crops pollination need was met by wild pollinators living within the farming landscape (Kevan and Phillips, 2001). Therefore, pollinators play paramount role in ecosystem through enhance the crop productivity and conserving biodiversity. On the above service of pollinators is priceless and natural endowment, but the society we live in, always categorized and measured in terms of financial value. In 2005, globally, insect pollination services economic value was estimated to be about €153 billion (Gallai *et al.,* 2009). The United Nations also reported that the economics of ecosystems and biodiversity by insect pollination was valued at £134 billion. The assisted pollination may have to be done when natural pollination is insufficient in order to reduce potential yield loss (Klein *et al.*, 2007).

Both wild and domesticated insect pollinators are declining and in parallel also those plants which are relying upon them are also declining gradually. Insect pollinators and other pollinator species are both declining at an alarming rate which is a threat to the existing plant pollination and their progeny. This depletion trend could damage dozens of commercially important crops. The declining phase of pollinator population and diversity is a serious concern to agricultural production, retention and conservation of biodiversity in many parts of the world, it also create scarcity of pollination services which have negative ecological and economic impacts that could significantly affect the plant diversity, wider ecosystem stability, crop productivity, food security and human welfare. The major causes of declining pollinators are worldwide due to habitat loss, habitat fragmentation, agrochemicals, monoculture, pathogens, enemies, exotic species, climate change and their interactions. Although the *Apis mellifera* has been widely accepted as one of the most important pollinators, however, population of honey bees are declining rapidly, causing global concern for pollination services (Biesmeijer *et al.,* 2006). On the other hand, some experts do not support the pollinator crisis hypothesis, but majority of the scientists recognized the importance of pollination services as also need to proliferate their population, diversity and support to continuous intensive research and monitoring worldwide. In the last two decades, there have been a number of examples observed globally; failure of crops directly due to the pollinator scarcity, *viz.* failure of pumpkins, cherries, alfalfa, blueberries, cashews and brazil nuts (Allen- Wardell *et al.,* 1998).

Due to ignorance of insect pollinators and their service are key constraints to the sustainability of contemporary agricultural practices. An initiative has been taken at the global level; the Convention on Biological Diversity has identified the importance of pollinators with the establishment of the International Initiative for the Conservation and Sustainable Use of Pollinators (International Pollinators Initiative-IPI) in 2000, facilitated and coordinated by FAO. Population is in increasing with decreasing rate day by day and food requirement have positive correlation with population, therefore food needs to be increase day by day for sustenance. Further cultivated land has been encroaching due to population increase with these activities such as construction of new roads, industrialization, new colonization etc. so land have negative correlation with population. The apparent link between pollinators and global food security and biodiversity, declining of managed and wild pollinators are of increasing concern. Increasing population has require more food therefore, each and every input play pivotal role towards productivity enhancement. Indian government has passed food security bill in parliament and in 2014 the population of our country is about 140 crores. To meet out the food requirement, pollination maximization and fertilized optimum ovules thereby enhance the productivity is one of the best methods, which shows great promises as a tool of 2nd green revolution as fifth input after land, labor, cost and capital. This can reduce increasing pressure on supply of agricultural land could significantly contribute to global environmental change.

Agro-ecosystem Characterization

Agro-ecosystem can be defined as a spatially and functionally coherent unit of agricultural activity, and includes the living and nonliving components involved in that unit as well as their interactions. This is a biological and natural resource system managed by humans for the primary purpose of producing food as well as other socially valuable nonfood goods and environmental services. The agro-ecosystem framework may be useful in the search for systems to sustain agriculture and natural resources in this region. Agro-ecosystem characterization as a natural vegetation, agricultural land use, crop diversity, artificial drainage, irrigation, net soil loss, and conservation practices show the extent to which resources have been modified to support agriculture.

Natural genetic diversity of plants has been denuded due to human interference by adoption of unscientific land use system. With rapid increase in human and livestock population and the rising demand of food, feed, fuel, fodder, fiber, timber and the other developmental activities, the farmers have been forced to exploit forestland and water resources at sub-optimal level in complete defiance of the inherent potential. Apart from this, the dwindling resources of soil, water, flora, fauna and increasing concern for environmental safety has drawn the attention of the planners and policy makers at regional as well national level.

In north east India, about 53 per cent of the reporting area is classified as forestland though there are large variations from minimum of 8 per cent in Meghalaya to a 91 per cent in Arunachal Pradesh. *Jhum* paddy is the dominant crop and is mixed with maize, millets, beans, tapioca, sweet potato, ginger, cotton, tobacco, chillies, sesamum and vegetables. Endowed with diverse climate, the NE region offers ideal agro-climatic conditions for cultivation of a wide variety of tropical and temperate fruits in its hills and valleys. The dominant horticultural crops of the region include pineapple, citrus, banana and areca nut. The region has a total area of 35.7 thousand ha under pineapple and 35.6 thousand ha under citrus. Orange is a unique crop of this region; however, average productivity is poor except in Assam and Meghalaya.

Status of Pollinators

Approximate more than 750,000 insect pollinators have been described (Grimaldi and Engel, 2005), possibly as many as 30 million more await discovery and formal description (Erwin, 2004). To ascertain the status of pollinator's species and their density involves consideration of a broader range of sources of information, including historical accounts, natural history collections, recently published observations, comparative analyses, asymptote and extinct. For some species, population data that are sufficient to inform an assessment of pollinator status simply do not exist. The assessment of the pollinator population's status in Medziphema, Nagaland 4 consecutively experiment was conducted and monitored. Population observation of 4year is not sufficient for any conclusion regarding population dynamics, nevertheless, its show trend of pollinators. However, on the basis of 4 year observation, survey, numbers of *Apis cerana* swarm colony, pollinator's species and their density observation on Sweat orange and Pigeonpea crop shows pollinators population are declining.

Depletion of insect pollinators in Nagaland due to destroy of pollinators dwelling by deforestation, *Jhum* cultivation, forest burn and tillage are harmful for pollinators and may cause a future problem to conserve the insect pollinators in the state. I corroborate bedrock of the Science report documented what appear to be a major decline in bees in England and the Netherlands, possibly a 30% loss in species richness since 1980, especially among specialist bees, and a corollary decline in wild plant species that require insect pollination (Biesmeijer *et al*., 2006).

Diversity of Pollinators

Maintaining the biological diversity of potential pollinators in agricultural ecosystems is not easy. However, it is necessary for the sustainability of agricultural crops that depend on these pollinators. In this context, bees play a prominent role in pollen transfer. A diverse group of pollinator species can provide better pollination services than a single species. The major pollinators belong to Hymenoptera order and out of this mostly belong to Apidae family, other animals, such as bats, birds, butterflies, moths, flies and beetles also play key roles in pollination. The diversity of pollinators and pollination systems are conspicuous. Most of the 25,000–30,000 species of bees (Hymenoptera: Apidae) are effective pollinators, and together with moths, flies, wasps, beetles and butterflies, make up the majority of pollinating species. Out of 100 principal crops of the worlds which are main source of food, only 15% are pollinated by domestic bees (mostly honey bees, bumble bees and alfalfa leafcutter bees), while at least 80% are pollinated by wild bees and other wildlife. Although wild bees generally shown a decline trend of diversity, abundance, and services with agricultural intensification (Tscharntke *et al*., 2005). The 25,000 different bee's species size and habit significantly differ to each other and differ accordingly in the plants they visit and pollinate, however the co- evolution of pollinators also depends upon related fauna. The diversity of fauna and their variability of food source also depend on the co- evolution of pollinators and their diversity. "Pollinators also play a vital role in sustaining wildlife and ecosystem health, both as part of the complex food chain and in the reproduction of plants. Evidently, measures are needed to ensure that our native and managed pollinator population is maintained and protected."

A more diverse community of wild pollinators can provide a greater amount of pollination services to a greater number of crops with greater stability. More diverse communities of pollinators in agricultural systems also have greater total abundances and rates of visitation to crop flowers (Pritchard, 2005; Chacoff and Aizen, 2006). Although many pollinator species that visit crops are generalists, different crop species nonetheless attract different, albeit partially overlapping, sets of pollinator species from the local species pool. Therefore, maintaining diverse pollinator communities locally is important for providing pollination services to a more diverse set of crops. Within a crop, a diverse group of pollinator species can provide better pollination services than a single species.

Habitats and Interactions of Pollinators (case studies)

Island biogeography theory predicts that species richness increases with habitat area and declines with isolation. The North east has rich biodiversity with gorgeous forest endowed by nature but presently due to suffering from deforestation, forest burning and *Jhum* cultivation, these malpractices make calamity for future. Destruction and fragmentation of natural and semi-natural habitats as well as land use intensification in agricultural landscapes have significant negative effects on honeybees and other pollinators (Kremen *et al.* 2007; Steffan-Dewenter and Westphal 2008; Tscharntke *et al.* 2005). The pollinators are currently under threat and this problem is arising from habitat fragmentation, destruction of foraging and nesting sites, their next generation (brood) and their stored food material in horrified way. Pollinators importance popularity in commence phase and the population density, diversity of honey bees and pollinators declining are also happening due to the habitat loss by forest eradication and *Jhum* cultivation. It has been observed that the population of both wild and managed pollinators is declining at alarming rates owing to alteration in their food and nesting habitats, shrinkage in natural ecosystems, i.e. forests and grassland ecosystems, pesticide poisoning, diseases and pests, excess harvesting of honey, introduction of K type of exotic pollinator species.

The North east is adventurous venture to domiciliation, augmentation and conservation of *Xylocopa* spp. through providing their natural habitat. Three species of Stingless bees (*Tetragonula iridipennis, Lophotrigona canifrons* and *Tetragonula laviceps*), Bumble bee, Megachilids, Blue band bee, Andrenids are efficient pollinators in natural habitats in these states. They play a vital role in sustaining the forest flora, epiphytic orchids. Stingless bees show wide range of foraging plants with short flight range of foraging behaviour than honey bees which make them the potential pollinators of future best suited to the needs of particular crops and habitat (Roubik 1995). Stingless bee's broad range of plant species visitor, they prefer small flowers, dense inflorescence, flowers with long corolla tubes. Stingless bees can be effectively used for green house pollination because of their limited flight range and the owner gets the maximum benefit. In some cases stingless bees show as nectar pollen robber due to removing nectar or pollen without pollinating the flowers of some crops. For example, stingless bees have been reported to damage the flowers of passion fruit and reduce the attraction of flower. In many countries where industrial farming is dominant, the natural population of solitary bees has declined due to destruction of their natural habitats. Bats pollinate wild bananas and North east is a predominated of wild banana state.

Climate changes affects biodiversity and changes in to natural ecosystems and affect climate. Climate change may lead to a sharp increase in rates of extinction. Thomas *et al.* (2004) studied five regions of the world, and predicted that if the present rate of climate change continues, 24% of species in these regions will be on their way to extinction by 2050. This study indicated that for many species, climate change poses a greater threat to their survival than the destruction of their natural habitat. "High altitudes are one of the habitats where it seems that climate change is having dramatic effects. According to them, the abundance of some flowers has changed, the timing of flowering has become earlier, particularly and the synchrony of plants and pollinators may be mismatch and plant may face scarcity of pollination whereas pollinators may feel floral dearth of floral rewards

Management for Pollination

Pollination management is tactics which accomplish pollination of flower, to enhance the quantity and quality of yield, by understanding of the particular crop's pollination requirements and pollination conditions. With the decline of both wild and domestic pollinator population density, increase the importance of best pollination management in key crops. The monoculture increases the importance of pollination management; it causes a diminutive food flow period and remaining part of the year pollinators suffering from scarcity of food mono-floral diet can reduce their immune system. In other hand a greater number of flowers in a short time required pollination with low abundance of pollinators. In this condition plant face pollination deficit and plant may not produce their optimum yield. In North east due to reach biodiversity pollinators does not depends upon list floral resource, they ate getting diversity of nectar and pollen while here major area cropping system is monotonous (paddy/ maize). Factors that cause the loss of pollinators include deforestation, forest burning, monoculture and *Jhum* cultivation in North east. Eradication of wild areas, loss of habitat and foraging plants corridors compel to migration of pollinators. In North east pollinators do not face pesticide hazardous due to very low consumption only government department are using DDT against mosquito is harmful for pollinators which is unfortunate. Organisms that are currently being used as pollinators in managed pollination are honey bees, bumblebees, stingless bee, alfalfa leafcutter bees, orchard mason bees, and carpenter bees in the world. Other species are expected to add to this list as this field develops.

The major crops of the world which have been managed pollination include apple, almonds, pears, plum cherry, cucumber coal crop seeds and onion seeds etc. Before planning and management of pollination needs to understand the agro- ecosystem, floral biology, true pollinators of particular crop, pollinator's abundance, habit and their habitats. The major challenges to know how manage wild pollinators for pollination service, exact taxonomic identification and habitat etc. Modify the farming practices in their fields which may benefit to pollinators, but the others management are also important viz. preserving wild habitat non-farmed land and field edges, to management of agricultural landscapes. Establish the landscape configurations that favour pollination services. Manage the cropping systems, flower-rich field margins, buffer zones and permanent hedgerows to ensure habitat and forage. In the field, pollinator-friendly practices include minimizing the use of farm chemicals, through organic production, integrated pest management, or finding alternatives to agrochemicals which is least toxic to pollinators.

The way farmers organize different land uses across their farms can influence pollination services. Avoid the monoculture, use multi and mixed cropping system, kitchen gardens and agro forestry systems also providing habitat and short dearth period on their farms for pollinators. Encouraged and provide the nest site for domiciliation to pollinators. Grow the shade trees and accommodation for domiciliation bee nest sites, e.g. by leaving standing dead trees and fallen branches undisturbed. To attract the pollinators and increasing their foraging rate and speed, particularly in the center of large plots for optimize quality and yield we may use semiochemical called SPLAT Bloom, which modifies the behavior of particular pollinators and inciting them to visit flowers in every portion of the field.

References

Allen-Wardell, G., Bernhardt, P., Bitner, R., Burquez, A., Buchmann, S., Cane, J., Cox, P.A., Dalton, V., Feinsinger, P., Ingram, M., Inouye, D., Jones, C.E., Kennedy, K., Kevan, P., Koopowitz, H., Medelllin, R., Medellin-Morales, S., Nabhan, G.P., Pavlik, B., Tepedino, V., Torchio, P. and Walker, S. 1998. The potential consequences of pollinator declines on the conservation of biodiversity and stability of food crop yields. Conservation Biology, 12(1):8-17.

Biesmeijer, J. C., Robert, S. P. M., Reemer, M., Ohlemuller, R., Edwards, M., Peeters, T., Schaffers, A. P., Potts, S. G., Kleukers, R., Thomas, C. D., Settele, J. and Kunin, W. E. 2006. Parallel declines in pollinators and insect-pollinated plants in Britain and the Netherlands. Science 313: 351- 354.

Chacoff, N. and Aizen, M. 2006. Edge effects on flower-visiting insects in grapefruit plantations bordering premontane subtropical forest. Journal of Applied Ecology, 43: 18- 27.

Erwin, T. L. 2004. The biodiversity question: How many species of terrestrial arthropods. Forest Canopies, 2nd edition, M.D.L. Lowman, and H.B. Rinker, eds. Burlington, Vt.: Elsevier Academic Press 259- 269.

FA0. 2007. Pollinators: Neglected biodiversity of importance to food and agriculture CGRFA-11/07/Inf.15. Food and Agriculture Organization of United Nations, Rome.

Gallai N, Salles JM, Settele J, Vaissiere BE (2009) Economic valuation of the vulnerability of world agriculture confronted with pollinator decline. Ecol Econ 68:810–821.

Grimaldi, D. and M. S. Engel. 2005. Evolution of the Insects. New York: Cambridge University Press.

Kearns, C. A., Inouye, D. W. and Waser, N. M. 1998. Endangered mutualisms: the conservation of plant-pollinator interactions. Annual Review of Ecology and Systematics, 29: 83–112.

Kevan, P. G., & Phillips, T. P. 2001. The economic impacts of pollinator declines: An approach to assessing the consequences. Conservation Ecology, 5: 8.

Klein, A. M., Vaissi, B.E., Are, J. H., Cane, I. Steffan-Dewenter, Cunningham, S. A., Kremen, C. and Tscharntke T. 2007. Importance of pollinators in changing landscapes for world crops. Proceedings of the Royal Society B: Biological Sciences. 274: 303- 313.

Kremen C., Williams N.M., Aizen M. A., Gemmill-Herren B. and LeBuhn G. 2007 Pollination and other ecosystem services produced by mobile organisms: a conceptual framework for the effects of land-use change. Ecol Lett. 10: 299– 314

Pritchard, K. 2005. The unseen costs of agricultural expansion across a rainforest landscape: Depauperate pollinator communities and reduced yield in isolated crops. Unpublished master's thesis, James Cook University, Queensland, Australia.

Roubik, D. W. 1995. Stingless bee colonies for pollination. Pollination of cultivated plants in the tropics. FAO agricultural service bulletin. FAO, Rome. 118: 150– 154.

Steffan-Dewenter, I., Westphal, C. 2008. The interplay of pollinator diversity, pollination services and landscape change. .J Appl. Ecol. 45: 737– 741.

Thomas, C. D., Cameron, A., Green, R. E., Bakkenes, M., Beaumont, L. J. and Collingham, Y. C. 2004. Extinction risk from climate change. Nature. 427: 145– 148.

Tscharntke, T., Klein, A. M., Kruess, A., Steffan-Dewenter, I. and Thies, C. 2005. Landscape perspectives on agricultural intensifi cation and biodiversity: Ecosystem service management. Ecology Letters. 8: 857– 874.

9

Effects of Climate Change on Insect Pollinators

Pollinators provide an essential ecosystem service, namely pollination, and it is an intricate relationship between plants and animals where reduction or loss of either will affect the survival of both. It has been established that pollinating agents are essential for survival and reproduction of several wild plant species (Kearns *et al.*, 1998). About 80% of the world's flowering plant species are dedicated to animal pollination, mostly by insects (FA0, 2007). Although more than 750,000 insect pollinators have been described (Grimaldi and Engel, 2005), possibly as many as 30 million more await discovery and formal description (Erwin, 2004). In 2005, globally, pollinators' pollination services economic value was estimated to be about €153 billion (Gallai *et al.,* 2009). The United Nations also reported that the economics of ecosystems and biodiversity by insect pollination was valued at £134 billion. The depletion of insect pollinators is due to pollinators' habitat destruction, heavy tillage, monoculture, injudicious use of pesticides and climate change and therefore, there is the need to create awareness on conservation and pollination management. Climate change is highly affected to species interaction as well as ecosystem ecology (van-der-Putten *et al.,* 2004; Sutherst *et al.,* 2007). Plant-pollinator interaction in both wild and cultivated plant species is under threat (Biesmeijer *et al.*, 2006).

Climate change is a significant change in the statistical distribution of weather patterns over periods ranging from decades to millions of years. It may be a change in average weather conditions, or in the distribution of weather around the average conditions (i.e., more or fewer extreme weather events). Its effect is already observed and projected in the context of natural ecosystems, biodiversity, human health and water resources. Plenty of studies already accomplished and concluded that climate change may be one of the biggest disturbance factors imposed on ecosystems towards less synchrony between historically interdependent species (Walther *et al.*, 2002 Parmesan, 2006). Climate change is one of the major concerns for pollination services (Memmott *et al.,* 2007; Hegland *et al.,* 2009; Schweiger *et al.,* 2010). The major effect of climate change is the increase in average global temperature which is directly associated with plant phenology. India is also highly threatened by climate change with experts warning that rising temperature will lead to more floods, siriasis, storm, rising sea levels and unpredictable crop production. In India temperature is rising at the rate of 0.4 °C every century of the year particularly during post monsoon and winter season (Ravindranath and Sathaye, 2002).

Phenology is the study of periodic events in the life cycles of animals or plants, as influenced by the environment (especially seasonal variations in temperature and precipitation). Climate change is altering the phenological response of plant and most of the pollinators may not be able to alter their life cycles and synchronization with altered pollinating timing. It is a direct threat to pollination services (Memmott *et al.,* 2007; Schweiger *et al.,* 2010; Hegland *et al.,* 2009). This asynchronization compels the insect pollinators towards migration, extinction; all these consequences create pollination deficits. Coope (1995) reported three possible scenarios for species' responses to large-scale climatic changes: adaptation to the new environment, emigration to another suitable area and extinction. Plant population may reduce due to pollination deficits and eventually they may become extinct and thereby, climate change affects the biodiversity. The timing of pollination is determined by climatic cues such as temperature and water availability (Cleland *et al*., 2007).

Effects of Climate Change on Pollination

Insect pollinators are valuable and limited resources (Delaplane and Mayer 2000). *Apis mellifera* is a most important cosmopolitan pollinator species for number of crops and their declining number in colonies is major concern and therefore, our reliance on single pollinators for such an important ecosystem service is risky. Some insect pollinators perpetuate in nature and their pollination credential ability has been recognized to be better than *Apis mellifera* for a particular crop. Biotic stress accompanied with climate change may cause population decline and lead to search for alternative pollinators for provident. The alfalfa leaf-cutter bee (*Megachile rotundata*) and alkali bee (*Nomia melanderi*) in alfalfa pollination (Cane 2002), mason bees (*Osmia* spp.) for pollination of orchards (Bosch and Kemp, 2002; Maccagnani *et al.,* 2003), bumblebees (*Bombus* spp.), *Amagilla* spp., play buzz pollination in solanaceous crops for pollination of crops requiring buzz pollination (Velthuis and van Doorn 2006) and stingless bee in green house condition are more effectual than honey bee due to short flight range.

The effect of climate change on pollinators depends upon their thermal tolerance and plasticity to temperature changes (Kjohl *et al*., 2011). The consequences of temperature due to climate change, induced alter in plant-pollinator interactions; Hegland *et al.* (2009) reported that timing of both plant flowering and pollinator activity seems to be strongly affected by temperature. Insects and plants may react differently to changed temperatures, creating temporal (phenological) and spatial (distributional) mismatches with severe demographic consequences for the species involved (FAO, 2011). Phenological asynchronization of plant and insect pollinators may cause pollination deficits thereby absence of pollinator/low diurnal abundance with low diversity; diminish insect visitation rate, and pollen deposition. This asynchronization may induce food scarcity for insect pollinators, some euryphagic insect pollinators may shift to other suitable flowering plants in same region or in another region due diminished floral rewards. A temporal mismatch can be affects to both plants and their pollinators. Climate change-induced mismatches in temporal (Hegland *et al.*, 2009) and spatial co-occurrence (Schweiger *et al.*, 2008), and morphological and physiological interdependencies of differently responding animal-pollinated plants and pollinators can potentially disrupt their interactions (Memmott *et al.*, 2007). However, Kudo *et al.* (2004) reported that early-flowering in Japan during a warm spring whereas bumble bee queen emergence unaffected by spring temperatures. Thus, direct temperature responses and the occurrence of mismatches in pollination interactions may vary among species and regions (Hegland *et al.,* 2009).

Effects of Climate Change on Flower

Flowering Phenology

Flowering phenology refers to the seasonal timing of flowering. Climate change affects the flowering phenology and it has negative impacts on plant pollination and pollinators due to asynchronization of flowering time and pollinator life cycle. Temperature, moisture and photoperiod are the three factors that affect the phenology of both plants and their pollinators (Partenen *et al.*, 1998). Due to global warming affects; timing of flowering has become earlier, floral abundance and asynchronization of plants-pollinators phenology. Across five studies from temperate regions of the USA and the UK, flowering commencement become advanced by 2- 6 days per 1 °C, with overall mean of 4 days. Asynchronization of plant-pollinators interactions is evident through recent changes in flowering phenologies (Fitter and fitter, 2002; Miller-Rushing *et al.*, 2007). Timing of flowering helps in maintaining reproductive isolation, actual floral calendar, overlapping blooming period and reducing the inter competition of pollinator. During these last decades, insect phenology showed a steeper advance than plant phenology and it cause asynchronization of plant- pollinators' interactions (Wielgolaski and Inouye 2003). Inouye *et al.*, 2003 reported that the climate change cause asynchrony of the flowering time and behaviour of pollinators such as butterflies, bumblebees and flies, then the intimate relationships between plants and pollinators that have co-evolved over thousands of years will be irreversibly altered (Inouye *et al.*, 2003).

Floral Biology

Floral biology refers to the structure, sexual system and morphological adaptations of the flowers in relation to the breeding system and pollination ecology. The effects of climate change on flower biology studies are very scanty. Some reports show rainfall, soil moisture and sunlight may alter the abundance and concentration of nectar in a flower and it could alter the foraging behaviour of pollinators. The honey bees' recruitment dance depends upon food source and food location; if food source and their volume will be altered, it will certainly affect bees' recruitment. It may affect the number of pollen transferred/ pollen deposition on stigma and visitation rate of the pollinators. Climate change leads to the frost period increase and erratic rain fall; frost leads to flower dropping in especially leguminous crops while uncertain rainfall leads to vegetative growth at the flowering time and escape the reproductive stage.

Drought negatively affects the plant growth and their productivity. Its impact on flower biology and pollination related studies are scanty, while some work accomplished on the impact of drought on crop yield excluded pollinators. In drought condition, flowers with few attractants were less attractive to pollinators (Galloway *et al.,* 2002; Pacini *et al,.* 2003; Mitchell *et al.,* 2004; Hegland and Totland 2005) and observed reductions in pollination levels, with poor seed quality and quantity (Philipp and Hansen, 2000; Kudo and Harder, 2005). Yield reduction under drought may also result from a decrease in pollen viability along with an increase in seed abortion rates, which have been identified as the most important factors affecting seed set (Melser and Klinkhamer, 2001).

Effects of Climate Change on Pollinators

Coope (1995) reported three possible scenarios for species' responses to large-scale climatic changes: adaptation to the new environment, emigration to another suitable area and extinction. I would like to mention one more important point here; climate change and evolution both are very slow process, it take long time, before extinction some K strategist insect pollinators (low reproductive with high survival) population decreases due to their low reproduction and higher death and emigration while survivors resist to acclimatize the climatic change with co-evolution. Therefore between emigration and extinction some pollinators' population become low and this resisting period may lead to evolution. Four possible steps may come in climatic changes:

1. Adaptation to the new environment
2. Emigration to another congenial habitat
3. Population depletion
4. Extinction

1. Adaptation to the New Environment

Adaptation is the co-evolutionary process of the environment and organisms for their survival in adverse environmental condition. The 2007, IPCC's report stated that "adaptation will be necessary to address impacts resulting from the warming which is already unavoidable due to past emissions." (IPCC 2007). As mentioned earlier, in the context of pollination, phenology is the change in timing of plant flowering and pollinators' behavior due to climatic change. These genetic changes in insect pollinators' populations have involved adaptation to the timing of seasonal events or to season length. In adverse climatic condition, K strategist insect may resist by their physiological change/ morphological change and adopt the new environment. The timing between seasonal events at low and high altitudes has negatively influenced migratory pollinators (Inouye *et al.*, 2000).

Usually, climate change impacts are easily noticed on phenotype or its physical features of insect pollinators' species. But there is a debate among scientists as to whether these phenotype changes reflect an adaptive genetic evolution or only phenotypic plasticity. Evolutionary adaptation is unlikely to be of major importance in the response of species to the climatic changes. Those insect pollinators who have short life cycle, multivoltine, cosmopolitan, eurythermal and euryphagic characters are considered to be highly adaptive pollinators.

2. Emigration to Another Congenial Habitat

Emigration is the one-way outward movement of the individuals from one place to another place. This may be due to inimical climatic condition viz. fierce temperature, flood, drought, food scarcity, etc. All these adverse circumstances obtrude by climate change. Each insect pollinator species require a certain range of temperature and beyond that they may not adopt but emigrate/ migrate to a congenial rage of temperature. The euryphagic pollinators have broad range of foraging plants, climate change may create temporary floral dearth while sternophagic insect pollinators have narrow range of foraging plants and climate change may create permanent floral dearth due to asynchronization of phenology. Temporary food scarcity compel to the pollinators migration (Singh *et al.*, 2007) whereas permanent floral dearth or asynchronization may force for emigration. In climate change, weather may not adopt the particular pollinator and they may not get suitable forage plants and due to these reasons they emigrate. Sternothermal and sternophagic insect pollinators have been badly affected from climate change. Emigration of insect pollinators may cause pollination deficits and ultimately loss of crop productivity.

3. Depletion of Population Density

Population density refers the number of individuals or population biomass per unit area. Depletion of population density depend upon 2 factors; high mortality rate in comparison to birth rate, high emigration rate than immigration. Climate change plays a major role for both factors; high mortality and high emigration and ultimately depletion of population density. The wild and domesticated insect pollinators are declining and in parallel, those plants that are relying upon them are also declining gradually. Insect pollinators and other pollinator species are declining at an alarming rate which is a threat to the existing plant pollination and the life of their progeny. Globally pollinators are declining due to so many causes, out of that climate change and their interactions play major role for depletion of population density. In the last two decades, number of examples reported globally; failure of crops (pumpkins, cherries, alfalfa, blueberries and cashews) directly due to the pollinator scarcity (Allen-Wardell *et al.,* 1998). Climate change has been affecting population of native pollinators; bumble bees and solitary bees and their population decreasing is in close connection to foraging plant extinction (Stefanescu *et al.*, 2003).

4. Extinction

Climate change threaten the insect pollinators by two way process, 1- obtrude the inimical weather condition, 2- asynchronization of plan pollinators phenology and create food scarcity to the pollinators. Both the above mentioned points conclude that climate change not only affects habitat, it affect the niche and make it unfavorable for pollinators. The population of both wild and managed pollinators is declining at alarming rates due to so many mankind interferences in natural ecosystem and climate change. The intricate relationships between plants and pollinators and their reduction may affect the survival of both mutualistic pollinators and their dependent plant species. Global extinction rate of species are accelerating at an alarming rate, it is estimated that 0.2- 0.3

per cent of all species are lost every year (Wilson, 1999). Global extinction of both mutualistic plant and pollinator species are increasing due to mankind activities; loss of habitat, introduction of exotic species and climate change (Kearns *et al*., 1998; Biesmeijer *et al*., 2006). Thomas *et al*., (2004) studied five regions of the world, and predicted that if the present rate of climate change continues, 24% of species in these regions will be on their way to extinction by 2050. In one example, variability in precipitation, linked to regional climate change, is considered as the cause of the extinction of two populations of checker spot butterflies in California (McLaughlin *et al*., 2002). Extinction of pollinators affects to the pollination ecosystem service and threatens the dependent plant species productivity and maintains their progeny.

Asynchrony Between Blooming and Pollinators' Activity

Usually, flowering time of plant species require a certain weather condition; it does not depend upon standard week or month. Insect pollinators also require congenial weather condition; they require a certain amount of temperature to complete their different stage of insects. Up to some extent, insect pollinators may adopt or acclimatize to the climatic condition. Of course, usually weather condition is synchronized with standard week/ month, but climate change make asynchronization and disturb the climatic relation with particular time of the year. Climate change increase the temperature and pre-pone the flowering time of plant species, while particular insect pollinators may not be available at the same time. This asynchrony between flowering time of plant and pollinators affect the crop pollination and decrease the crop productivity.

References

Allen-Wardell, G., Bernhardt, P., Bitner, R., Burquez, A., Buchmann, S., Cane, J., Cox, P.A., Dalton, V., Feinsinger, P., Ingram, M., Inouye, D., Jones, C.E., Kennedy, K., Kevan, P., Koopowitz, H., Medelllin, R., Medellin-Morales, S., Nabhan, G.P., Pavlik, B., Tepedino, V., Torchio, P. and Walker, S. 1998. The potential consequences of pollinator declines on the conservation of biodiversity and stability of food crop yields. Conservation Biology, 12(1):8-17.

Biesmeijer, J.C., Roberts, S.P. M., Reemer, M., Ohlemuller, R., Edwards, M., Peeters, T., Schaffers, A.P., Potts, S.G., Kleukers, R., Thomas, C.D., Settele, J. and Kunin, W.E. 2006. Parallel declines in pollinators and insect-pollinated plants in Britain and the Netherlands. Science, 313:351-354.

Bosch, J. and Kemp, W. P. 2002. Developing and establishing bee species as crop pollinators: the example of Osmia spp. (Hymenoptera : Megachilidae) and fruit trees. Bulletin of Entomological Research, 92: 3-16.

Cane, J.H. 2002. Pollinating bees (Hymenoptera : Apiformes) of US alfalfa compared for rates of pod and seed set. Journal of Economic Entomology, 95: 22-27.

Coope, G.R. 1995. Insect faunas in ice age environments: why so little extinction? In J. Lawton R. May, eds. Extinction rates, pp 55-74. Oxford, UK, Oxford Univ. Press.

Delaplane, K.S. and Mayer, D.F. 2000. Crop pollination by bees. New York, CABI.

Erwin, T. L. 2004. The biodiversity question: How many species of terrestrial arthropods. Forest Canopies, 2nd edition, M.D.L. Lowman, and H.B. Rinker, eds. Burlington, Vt.: Elsevier Academic Press 259- 269.

FA0. 2007. Pollinators: Neglected biodiversity of importance to food and agriculture CGRFA-11/07/Inf.15. Food and Agriculture Organization of United Nations, Rome.

Kjohl, M., Nielsen, A., Stenseth, N.C., 2011. Potential Effects of Climate Change on Crop Pollination. FAO, Rome. http://typo3.fao.org/fileadmin/templates/agphome/documents /Biodiversity-pollination/Climate_Pollination_17_web__ 2_.pdfFitter, A.H. and Fitter, R.S.R. 2002. Rapid changes in flowering time in British plants. Science. 296:1689- 1691.

Gallai, N., Vaissiere, B. E., Potts, S. G. and Salles, J. 2009. Assessing the monetary value of global crop pollination services. Oxford University Press.

Galloway, L.F., Cirigliano, T. and Gremski, K. 2002. The contribution of display size and dichogamy to potential geitonogamy in Campanula americana. International Journal of Plant Science, 163: 133-139.

Grimaldi, D. and M. S. Engel. 2005. Evolution of the Insects. New York: Cambridge University Press.

Hegland, S.J. and Totland, O. 2005. Relationships between species' floral traits and pollinator visitation in a temperate grassland. Oecologia, 145: 586-594.

Hegland, S.J., Nielsen, A., Lazaro, A., Bjerknes, A.L. and Totland, O. 2009. How does climate warming affect plant-pollinator interactions? Ecology Letters, 12: 184-195.

Inouye DW (2000) The ecological and Evolutionary significance of frost in the context of climate change. Ecology Letters, 3(5):457- 463.

Inouye, D. W., Saavedra, F., and Lee-Yang, W. 2003. Environmental influences on the phenology and abundance of flowering by Androsace septentrionalis L. (Primulaceae). American Journal of Botany, 90(6): 905- 910.

Kearns, C.A., Inouye, D.W. and Waser, N.M. 1998. Endangered mutualisms: the conservation of plant-pollinator interactions. Annual Review of Ecology Evolution and Systematics, 29: 83-112.

Kudo, G. and Harder, L.D. 2005. Floral and inflorescence effects on variation in pollen removal and seed production among six legume species. Functional Ecology, 19: 245- 254.

Kudo, G., Nishikawa, Y., Kasagi, T. and Kosuge, S. 2004. Does seed production of spring ephemerals decrease when spring comes early? Ecological Research, 19: 255-259.

Maccagnani, B., Ladurner, E., Santi, F. and Burgio, G. 2003. Osmia cornuta (Hymenoptera, Megachilidae) as a pollinator of pear (Pyrus communis): fruit- and seed-set. Apidologie, 34: 207-216.

Mclaughlin, J. F., Hellmann, J. J., Boggs, C. L. & Ehrlich, P. R. (2002) Climate change hastens population extinctions. Proceeding of National Academic of Sciences, 99(9):6070–6074.

Melser, C. and Klinkhamer, P.G.L. 2001. Selective seed abortion increases offspring survival in Cynoglossum officinale (Boraginaceae). American Journal of Botany, 88: 1033-1040.

Memmott, J., Craze, P.G., Waser, N.M. and Price, M.V. 2007. Global warming and the disruption of plant-pollinator interactions. Ecology Letters, 10: 710- 717.

Miller-Rushing, A.J., Katsuki, T., Primack, R. B., Ishii, Y., Lee, S. D. and Higuchi, H. 2007. Impact of global warming on a group of related species and their hybrids: cherry tree (Rosaceae) flowering at Mt. Takao, Japan. American Journal of Botany, 94:1470- 1478.

Mitchell, R.J., Karron, J.D., Holmquist, K.G. and Bell, J.M. 2004. The influence of Mimulus ringens floral display size on pollinator visitation patterns. Functional Ecology, 18: 116-124.

Pacini, E., Nepi, M. and Vesprini, J.L. 2003. Nectar biodiversity: a short review. Plant Systematics Evolution, 238: 7- 21.

Parmesan, C. 2006. Ecological and evolutionary responses to recent climate change. Annual Review of Ecology Evolution and Systematics, 37: 637- 669.

Partanen, J., Koski, V. and Hanninen, H. 1998. Effects of photoperiod and temperature on the timing of bud burst in Norway spruce (Picea abies). Tree Physiology, 18: 811- 816.

Philipp, M. and Hansen, T. 2000. The influence of plant and corolla size on pollen deposition and seed set in Geranium sanguineum (Geraniaceae). Nordic Journal of Botany, 20: 129-140.

Ravindarnath, N. H. and Sathaye, T. H. 2002. Climate change and developing countries (Advances in Global Change Research), Kluwer Academic Publishers, Dordrecht.

Schweiger, O., Settele, J., Kudrna, O., Klotz, S. and Kuhn, I. 2008. Climate change can cause spatial mismatch of trophic interacting species. Ecology, 89: 3472- 3479.

Schweiger, O., Biesmeijer, J.C., Bommarco, R., Hickler, T., Hulme, P., Klotz, S., Kuhn, I., Moora, M., Nielsen, A., Ohlemuller, R., Petanidou, T., Potts, S.G., Pysek, P., Stout, J. C., Sykes, M., Tscheulin, T., Vila, M.,Wather, G.R. and Westphal, C. 2010. Multiple stressors on biotic interactions: how climate change and alien species interact to affect pollination. Biological Review, 85: 777- 795.

Singh, R. P. Singh, Singh, A. K. and Singh, R. P. 2007. The effect of the availability of bee forage plant and environmental conditions on the nesting of Apis dorsata Fabr. Journals of Apicultural Research, 46 (4):276- 281.

Stefanescu, C., Penuelas, J. and Fililla, I. 2003. Effects of climate change on the phenology of butterflies in the northwest Mediterranean basin. Global Change Biology, 9:1494- 1506.

Sutherst, R.W., Maywald, G.F. and Bourne, A.S. 2007. Including species interactions in risk assessments for global change. Global Change Biology, 13: 1843-1859.

Thomas, C.D. , Cameron, A., Green, R.E., Bakkenes, M., Beaumont, L.J., Collingham, Y.C., Erasmus, B.F.N., de-Siqueira, M.F., Grainger, A., Hannah, L., Hughes, L., Huntley, B., van-Jaarsveld, A.S., Midgley, G. F., Miles, L., Ortega-Huerta, M. A. , Peterson, A.T., Phillips, O.L. and Williams, S.E. 2004. Extinction risk from climate change. Nature, 427:145- 148.

van der Putten, W.H., de Ruiter, P.C., Bezemer, T.M., Harvey, J.A., Wassen, M. and Wolters, V. 2004. Trophic interactions in a changing world. Basic and Applied Ecology, 5: 487- 494.

Velthuis, H.H.W. and van Doorn, A. 2006. A century of advances in bumblebee domestication and the economic and environmental aspects of its commercialization for pollination. Apidologie, 37: 421- 451.

Walther, G.R., Post, E., Convey, P., Menzel, A., Parmesan, C., Beebee, T.J.C., Fromentin, J.C., Hoegh-Guldberg, O. and Bairlein, F. 2002. Ecological responses to recent climate change. Nature, 416: 389- 395.

Wielgolaski, F. E. and Inouye, D. W. 2003. Phenology of high-latitude climates. In: Schwartz MD (ed) Phenology: an integrative environmental science. Kluwer Academic Publishers, Dordrecht, pp 175- 194.

Wilson, E.O. 1988. The current state of biodiversity. In: Wilson EO, Peter FM (eds) Biodiversity. National Academic Press, Washington DC, pp 3-18

10

Future Prospect of Agricultural Pollination

Approximately 80 per cent of all flowering plant species are specialized for pollination by animals and mostly by insects. These crops account for 65 per cent of global food production, still leaving as much as 35 per cent depending on pollinating animals (Klein *et al.,* 2007). These food crops include delicious products such as many of our tree fruits (e.g., apples, cherries, mangos, and avocados) and nuts (e.g., almonds, peanuts, macadamia nuts, and cashew nut), squashes, cucumbers, melons, citrus, and berries. In addition, many of oil seed crops benefit from bee pollination, such as mustard, coconut, canola, safflower, and many more.

The honey bee (*Apis mellifera*) is the most commonly used insect in agricultural crops as managed pollinator in North America and Europe, Japan, China, India and many other parts of the world as well. Farmers in the United States rent more than 2.5 million honey bee colonies every year for fulfillment of pollination deficit but today, the honey bee is threatened by colony collapse disorder (CCD). The demand for honey bees in California almond orchards rapidly increased due to increasing acreage, while on the other side, bee colonies decreased due to CCD, but that occurred during a time when honey bee colonies were not easy to procure and as a result, the cost of renting bees shot up from US$50 per hive in 2003 to US$140 per hive in 2006 (Sumner and Boriss, 2006). The rental charge increased about 3 times within 4 years.

In addition, not all crops are well pollinated by honey bees. For example, tomato, brinjal, etc. require buzz pollination, where honey bees cannot accomplish pollination; alfalfa flowers are not properly pollinated by honey bees, pigeonpea flower are not pollinated properly by honey bees and honey bees do not work well in greenhouses. Fortunately, nature provides endowments of diverse pollinators to the ecosystem for important service; pollination. Approximately 16,000 species of bees are known in the world (Michener, 2000) and only a few species have been managed for crop pollinators. These include honey bees, bumble bees, leafcutter bees, mason bees, alkali bees, carpenter bees and stingless bees. There exists a vast potential to manage many more species of bees and to broaden the use of those that are already being used. However, the management of non-*Apis* pollinators poses its own set of problems. Nesting materials, shelters, and rearing techniques have been well developed for the alfalfa leafcutter bee and are commercially available but methods for propagating other important pollinators, such as blue orchard bees and bumble bees are less developed. Furthermore, these bees have their own set of disease and parasite problems that needs to be addressed with concern.

The enormous diversity of wild bees and other insect pollinators have been unexplored, in regard to the plants pollination. The taxonomic identification and systematic knowledge of many important pollinator groups is inadequate due to lack of taxonomist. In addition, we lack an understanding of native bee biology, habitat, food preference and pests of bees. Some of our native bees are threatened by a shrinking natural habitat and a lack of information about their biology, bionomics and natural population regulation inhibits our ability to rely on their populations for pollination. More information is needed to help in augmentation and conservation of native bees.

The complete genome of the honey bee was sequenced and reported in *Nature* (Honey bee Genome Sequencing Consortium, 2006), and the genomes of two major honey bee pathogens were sequenced, *Ascosphaera apis* and *Paenibacillus larvae* (Qin *et al.*, 2006). At about the same time, a new phylogenetic analysis that included gene sequence data was used to reevaluate our previous conceptions of the evolution of bees (Danforth *et al.*, 2006). Earlier in the year, a *Science* report documented in what appeared to be a major decline in bees in England and the Netherlands, possibly a 30 per cent loss in species richness since 1980, especially among specialist bees, and a corollary decline in wild plant species that require insect pollination (Biesmeijer *et al.*, 2006).

As the world population is growing rapidly, the demand of food production will be double in the next twenty-seven years. The cultivated land is limited and it has been encroached by new construction, and therefore, the only one way to overcome this challenge is to increase the productivity. Undoubtedly, agricultural production, productivity and agro-ecosystem diversity are threatened by declining populations of pollinators and their extinction. Many pollinator population densities have been reduced and the available population are unable to fulfil the required pollination service of plants. So therefore, plant production, productivity and diversity of dependent crops are in peril. The declining trend of pollinators' density and diversity are of major concern for ecosystem service to sustain the crop productivity and plant biodiversity. Augmentation and conservation of pollinators to provide sufficient pollination of crops is also one of the best methods to enhance the productivity which may be expected to be a tool of 2nd green revolution as fifth input after land, labor, cost and capital.

In India, National Commission on Agriculture in 1976, recommended the starting of a project entitled "All India Coordinated project on Honey bee Research & Training". Honey bee pollination is 13-18 times more lucrative than bee hive products, since 80 per cent of crops rely on pollinators for their pollination and many more aspects. As per the importance of pollinators, the objectives were refined and the title of this project was re-designated as All India Coordinated Research Project on Honey Bees & Pollinators in July 2007. The NAS report, Status of Pollinators in North America, highlights areas of concern that directly relate to providing adequate pollinators for agriculture and the environment. The Centre for Pollination Studies (CPS) has been established at the University of Calcutta through the Darwin Initiative of the Department for Food and Rural Affairs (DEFRA) in partnership with the Government of India's Department of Science & Technology. The CPS focuses on pollination systems with an expectation that as the centre grows, the research will extend into other areas that contribute to the development of sustainable farming.

Before World War II, the manner of farming and agriculture was often "agrarianism" and in that procedure, crops were grown in subsistence phase. The term "green revolution", coined by Dr. William Gaud in 1968, is meant for enhancement of the photosynthetic activity of the chlorophyll leading to more production. This evolution utilizes solar energy, water management and several nutrients. Now, the term "evergreen revolution" was coined by the eminent scientist Dr. M. S. Swaminathan and it refers to enhancement of the productivity in perpetuity without associated ecological harm. After 2007, augmentation of pollinators became a 5th input to enhance the productivity which, as expected, may be a tool of 2nd evergreen revolution as fifth input after land, labor, cost and capital. This increasing pressure on supply of agricultural land could significantly contribute to global environmental change. To gain the attention of the general public and business corporations, conservationists' arguments are often made in the context of the economic value of wildlife to society. Mariola's (2005) argument for a return to agrarianism as a basis for farm preservation is reminiscent of the wildlife conservation. A land ethic then, reflects the existence of an ecological conscience, and this in turn reflects a conviction of individual responsibility for the health of the land.

For providence, it is our need to conserve and, as much as possible, proliferate the pollinator population till asymptote cross the boundaries between agrarianism and a land ethic, despite the fact that the values and goals of farmers and biodiversity conservationists are not always in synchrony with each other. For example, many species of bees are rare or specialized and have no known utilitarian value to agriculture. Conservation efforts generally target the preservation of rare or endangered species and successful conservation often calls for the sacrifice of some human activity. These activities can include agricultural practices such as use of pesticide, tillage, or other land uses. It is true that bees produce direct benefit for humans by pollinating agricultural crops and gardens but they are also important for maintaining a balanced ecosystem, on which we are dependent and from which we can learn social values. These small, unobtrusive creatures are symbolic of our interconnection with nature and highlight the ethical responsibility we have toward maintaining the integrity of the natural world.

References

Biesmeijer, J. C., Roberts, S. P. M., Reemer, M., Ohlemuller, R., Edwards, M., Peeters, T., Schaffers, A. P., Potts, S.G., Kleukers, R., Thomas, C. D., Settele, J. and Kunin, W. E. 2006. Parallel declines in pollinators and insect-pollinated plants in Britain and the Netherlands. Science, 313:351- 354.

Danforth, B. N. S., Sipes, S., Fang, J. and Brady, S. G. 2006. The history of early bee diversifi cation based on fi ve genes plus morphology. Proceedings of the National Academy of Sciences of the USA, 103, pp. 15118- 15123.

Honey bee Genome Sequencing Consortium. 2006. Insights into social insects from the genome of the honey bee Apis mellifera. Nature, 443, 931- 949.

Klein, A-M., Vaissi-Are, B. E., Cane, J. H., Steffan-Dewenter, I., Cunningham, S.A, Kremen, C. and Tscharntke, T. 2007. Importance of pollinators in changing landscapes for world crops. Proceedings of the Royal Society B: Biological Sciences 274: 303- 313.

Mariola, M. J. 2005. Losing ground: Farmland preservation, economic utilitarianism, and the erosion of the agrarian ideal. Agriculture and Human Values, 22: 209- 223.

Michener, C. D. 2000. The bees of the world. The John Hopkins University Press, Baltimore/ London, 913 pp.

Qin, X., Evans, J. D., Aronstein, K. A., Murray, K. D., and Weinstock, G. M. 2006. Genome sequences of the honey bee pathogens Paenibacillus larvae and Ascosphaera apis. Insect Molecular Biology, 15, 715- 718.

Sumner, D. A. and Boriss, H. 2006. Bee-conomics and the leap in pollination fees. Giannini Foundation of Agricultural Economics, 9: 9- 11.

Suggested Readings

Abrol, D.P. 2009. Bees and beekeeping in India, 2nd edn. Kalyani Publishers, Ludhiana, 720 p.

Abrol, D.P., 2012. Pollination biology- biodiversity conservation and agricultural productivity. Springer Dordrecht Heidelberg London New York, 787 pp.

Aizen, M. and Harder, L.D. 2009. The global stock of domesticated honey bees is growing slower than agricultural demand for pollination. Current Biology, 19:915-918.

Aizen, M., Garibaldi, L., Cunningham, S. and Klein, A. 2008. Long-term global trends in crop yield and production reveal no current pollination shortage but increasing pollinator dependency. Current Biology, 18:1572-1575.

Allen-Wardell, G., Bernhardt, P., Bitner, R., Burquez, A., Buchmann, S., Cane, J., Cox, P.A., Dalton, V., Feinsinger, P., Ingram, M., Inouye, D., Jones, C.E., Kennedy, K., Kevan, P., Koopowitz, H., Medelllin, R., Medellin-Morales, S., Nabhan, G.P., Pavlik, B., Tepedino, V., Torchio, P. and Walker, S. 1998. The potential consequences of pollinator declines on the conservation of biodiversity and stability of food crop yields. Conservation Biology, 12(1):8-17.

Banda, H.J. and Paxton, R.J. 1991. Pollination of greenhouse tomatoes by bees. Acta Horticulture, 288:194-198.

Biesmeijer, J.C., Roberts, S.P. M., Reemer, M., Ohlemuller, R., Edwards, M., Peeters, T., Schaffers, A.P., Potts, S.G., Kleukers, R., Thomas, C.D., Settele, J. and Kunin, W.E. 2006. Parallel declines in pollinators and insect-pollinated plants in Britain and the Netherlands. Science, 313:351-354.

Brown, M.J.F. and Paxton, R.J. 2009. The conservation of bees: a global perspective. Apidologie, 40:410-416.

Buchmann, S. L. and Ascher J. S. 2005. The plight of pollinating bees. Bee world, 86:71-74.

Buchmann, S.E. 1983. Buzz pollination in angiosperms. Handbook of experimental pollination biology. Van Nostrand Reinhold, New York, pp 73-113.

Buchmann, S.E. and Nabhan, G.P. 1996. The forgotten pollinators. Island Press, Washington, DC.

Burd, M. 1994. Bateman's principle and plant reproduction: the role of pollen limitation in fruit and seed set. Botanical Review, 60: 83-139.

Carre, G. et al. 2009. Landscape context and habitat type as drivers of bee diversity in European annual crops. Agriculture Ecosystem Environment, 133: 40- 47.

Chacoff, N. and Aizen, M. 2006. Edge effects on flower-visiting insects in grapefruit plantations bordering premontane subtropical forest. Journal of Applied Ecology, 43: 18-27.

Dafni, A. 1998. The threat of Bombus terrestris spread. Bee World, 79:113-114.

Daily, G. C., Alexander, S., Ehrlich, P.R., Goulder, L., Lubchenco, J., Matson, P.A., Mooney, H.A., Postel, S., Schneider, S.H., Tilman, D. and Woodwell, G.M. 1997. Ecosystem services: benefits supplied to human societies by natural ecosystems. Issues in Ecology: 1-16.

Danforth, B.N.S., Sipes, S., Fang, J. and Brady, S.G. 2006. The history of early bee diversifi cation based on fi ve genes plus morphology. Proceedings of the National Academy of Sciences of the USA, 103, pp. 15118-15123.

Delaplane, K.S., Mayer, D.F., 2000. Crop Pollination by Bees. CABI Publishing, UK.

Erwin, T. L. 2004. The biodiversity question: How many species of terrestrial arthropods. Forest Canopies, 2nd edition, M.D.L. Lowman, and H.B. Rinker, eds. Burlington, Vt.: Elsevier Academic Press 259- 269.

FA0. 2007. Pollinators: Neglected biodiversity of importance to food and agriculture CGRFA-11/07/Inf.15. Food and Agriculture Organization of United Nations, Rome.

FAO. 1993. Harvesting nature's diversity. FAO, Rome.

Free, J.B. 1993. Insect pollination of crops, 2nd edn. Academic, London.

Freitas, B.M. and Oliveira-Filho, J.H. 2001. Criacao racional de mamangavas para polinizacao em areas agrícolas. BNB, Fortaleza.

Freitas, B.M., Otavio, J., Pereira, P. and Fortaleza, C.E. 2004. Solitary bees conservation, rearing and management for pollination. A contribution to the International Workshop on Solitary Bees and Their Role in Pollination, Beberibe, Ceara, Brazil, in Aprl., 282 p.

Gabriel, D. and Tscharntke, T. 2007. Insect pollinated plants benefit from organic farming. Agriculture Ecosystem Environment, 118: 43- 48.

Gallai, N., Vaissie`re, B. E., Potts, S. G. and Salles, J. 2009. Assessing the monetary value of global crop pollination services. Oxford University Press.

Ghazoul, J. 2005. Buzziness as usual? Questioning the global pollination crisis. Trends of Ecology Evolution, 20:367-373.

Goulson, D. et al 2008. Decline and conservation of bumble bees. Annual Review Entomology, 53:191-208.

Grimaldi, D. and M. S. Engel. 2005. Evolution of the Insects. New York: Cambridge University Press.

Holzschuh, A. et al. 2008. Agricultural landscapes with organic crops support higher pollinator diversity. Oikos, 117:354-361.

Honey bee Genome Sequencing Consortium. 2006. Insights into social insects from the genome of the honey bee Apis mellifera. Nature, 443, 931-949.

Jaffee, R. 2010. Estimating the density of honey bee colonies across their natural range to fill the gap in pollinator decline censuses. Conservation Biology, 24:583-593

Jones, D.L. 1975. The pollination of Microtis parviflora R. Br. Annals of Botany, 39:585-589.

Kearns, C.A., Inouye, D.W. and Waser, N.M. 1998. Endangered mutualisms: the conservation of plant- pollinator interactions. Annual Review Ecology Systematic, 29:83-112.

Kerr, E.A. and Kribs, W. 1955. Electric vibrator as an aid in greenhouse tomato production. Queensland Journal of Agriculture Science, 2:157-169.

Kevan, P.G. 1975. Pollination and environmental conservation. Environment Conservation, 2:293-298.

Kevan, P.G. 1999. Pollinators as bioindicators of the state of the environment: species, activity and diversity. Agriculture Ecosystem Environment, 74:373-393.

Kevan, P.G. and Plowright, R.C. 1995. Impact of pesticides on forest pollination. In: Armstrong JA, Ives WGH (eds) Forest insect pests in Canada. pp 607-618.

Kevan, P.G., Straver, W.A., Offer, M. and Laverty, T.M. 1991. Pollination of greenhouse tomatoes by bumblebees in Ontario. Proceeding of Entomology Society, Ontario 122:15-19.

Khan, K.A., Ahmad, K.J., Razzaq, A., Shafiqe, M., Abbasi, K.H., Saleem, M. and Ullah, M.A. (2012). Pollination Effect of Honey Bees, Apis mellifera L. (Hymenoptera: Apidae) on Apple Fruit Development and its Weight. Persian Gulf Crop Protection, 1(2): 1-5.

Klein, A-M., Vaissi-Are, B. E., Cane, J.H., Steffan-Dewenter, I., Cunningham, S.A, Kremen, C. and Tscharntke, T. 2007. Importance of pollinators in changing landscapes for world crops. Proceedings of the Royal Society B: Biological Sciences 274: 303-313.

Klein, A-M., Vaissiere, B.E., Cane, J.H., Steffan-Dewenter, I., Cunningham, S.A. et al. 2007. Importance of pollinators in changing landscapes for world crops. Proceedings of the Royal Society B: Biological Science, 274: 303- 313.

Kremen, C. et al. 2002. Crop pollination from native bees at risk from agricultural intensification. Proceeding National Academic Science, USA 99:16812-16816.

Kremen, C., Williams, N.M., Aizen, M.A., Gemmill-Herren, B., LeBuhn, G., Minckley, R., Packer, L., Potts, S.G., Roulston, T., Steffan-Dewenter, I., Vazquez, D.P., Winfree, R., Adams, L., Crone, E.E., Greenleaf, S.S., Keitt, T.H., Klein, A-M, Regetz, J. and Ricketts, T.H. 2007. Pollination and other ecosystem services produced by mobile organisms: a conceptual framework for the effects of land-use change. Ecology Letter, 10:299-314.

Larson, B.M.H., Kevan, P.G. and Inouye, D.W. 2001. Flies and flowers: taxonomic diversity of anthophiles and pollinators. Canadian Entomology, 133:439-465.

Lavergne, S. et al. 2006. Fingerprints of environmental change on the rare Mediterranean flora: a 115-year study. Global Change Biology, 12:1466-1478.

Linsley, E.G. 1958. The ecology of solitary bees. Hilgardia, 27:543-599.

Mariola, M. J. 2005. Losing ground: Farmland preservation, economic utilitarianism, and the erosion of the agrarian ideal. Agriculture and Human Values, 22: 209-223.

Martins, D., Gemmill, B. and Eardley, C. 2003. Plan of action of the African pollinator initiative. African Pollinator Initiative, Nairobi.

McGregor, S.E. 1976. Insect pollination of cultivated crop plants. United States Department of Agriculture (USDA), Washington DC.

Michener, C. D. 2007. The bees of the world, 2nd Edn. The Johns Hopkins University Press, Baltimore.

Moffett, J.O., Morton, H.L. and Macdonald, R.H. 1972. Toxicity of some herbicidal sprays to honey bees. Journal of Economic Entomology, 65:32-36.

Moritz, R.F.A. et al. 2007. The size of wild honey bee populations (Apis mellifera) and its implications for the conservation of honey bees. Journal of Insect Conservation, 11:391-397.

Morse, R.A. 1978. Honey bee pests, predators and diseases. Cornell University Press, Ithaca, 430 pp.

Muller, H. 1883. The fertilization of flowers. In:Translated and edited by Thompson DW. Macmillan, London, 669 pp.

Nabhan, G. P. and Buchmann, S. 1997. Services provided by pollinators. In G. C. Daily (Ed.), Nature's services: Societal dependence on natural ecosystems (133-150). Washington, DC: Island Press.

National Research Council. 2007. Status of pollinators in North America. National Academies Press, Washington, DC: 307 p.

Natural Research Council (2006) Status of pollinators in North America. National Academic Press, Washington, DC.

Partap, U. 2001. Warning signals from the Apple Valleys. Video Film, 31 minutes and 7 seconds, VHS Format, ICIMOD, Kathmandu.

Partap, U. and Partap, T. 1997. Managed crop pollination: the missing dimension of mountain agricultural productivity. Mountain Farming Systems' Discussion Paper Series No. MFS 97/1, ICIMOD, Kathmandu.

Partap, U. and Partap, T. 2002. Warning signals from the apple valleys of the HKH: Productivity concerns and pollination problems. ICIMOD, Kathmandu, 124 p.

Petanidou, T., Ellis, W.N., 1993. Pollinating fauna of a phryganic ecosystem: composition and diversity. Biodiversity Letters, 1, 9-22.

Pimentel, D., Acquay, H., Biltonen, M., Rice, P., Silva, M., Nelson, J., Lipner, V., Giordano, S., Horowitz, A., and D'Amore, M., 1992a, Environmental and economic costs of pesticide use, BioScience 42:750-760.

Pimentel, D., Stachow, U., Takacs, D.A., Brubaker, H.W., Dumas, A.R., Meaney, J.J., O'Neil, J., Onsi, D.E. and Corzilius, D.B. 1992b. Conserving biological diversity in agricultural/ forestry systems. BioScience, 42:354-362.

Potts, S.G., Roberts, S.P.M.., Dean, R., Marris, G., Brown, M.A., Jones, H.R., Neumann, P. and Settele, J. 2010. Declines of managed honey bees and beekeepers. European Journal of Apicultural Research, 49(1):15-22.

Potts, S.G., Vulliamy, B., Dafni, A., Ne'eman, G., O'Toole, C., Roberts, S., Willmer, P.G., 2003. Response of plant-pollinator communities following fire: changes in diversity, abundance and reward structure. Oikos, 101, 103-112.

Pritchard, K. 2005. The unseen costs of agricultural expansion across a rainforest landscape: Depauperate pollinator communities and reduced yield in isolated crops. Unpublished master's thesis, James Cook University, Queensland, Australia.

Proctor, M. and Yeo, P. 1973. The pollination of flowers. Collins, London, 418 pp. Ann Miss Bot Gard, 61:770-80.

Qin, X., Evans, J. D., Aronstein, K. A., Murray, K. D., and Weinstock, G. M. 2006. Genome sequences of the honey bee pathogens Paenibacillus larvae and Ascosphaera apis. Insect Molecular Biology, 15, 715-718.

Rana, B.S., Garg, Paul, I.D., Khurana, S.M., Verma, L.R. and Agrawal, H.O. 1987. Thai sacbrood virus of honey bees (Apis cerana indica F) in the North West Himalayas. Indian Journal of Virology, 2:127-131.

Roubik, D.W. 1989. Ecology and natural history of tropical bees. Cambridge University Press, New York.

Roubik, D.W. 1995. Pollination of cultivated plants in the tropics. Food and Agricultural Organization of the United Nations, Rome, Bulletin 118.

Sammataro, D. et al. 2000. Parasitic mites of honey bees: life history, implications, and impact. Annual Review Entomology, 45:519-548.

Sciencedaily. 2008. Economic Value Of Insect Pollination Worldwide Estimated At U.S. $217 Billion.

Steffan-Dewenter, I., Potts, S.G. and Packer, L. 2005. Pollinator diversity and crop pollination services are at risk. Trends of Ecology Evolution, 20:651-652.

Stephen, W.P. 1955. Alfalfa pollination in Manitoba. Journal of Economic Entomology, 48:543-548.

Sumner, D. A. and Boriss, H. 2006. Bee-conomics and the leap in pollination fees. Giannini Foundation of Agricultural Economics, 9: 9-11.

Sutton, S.L. and Collins, N.M. 1991. Insects and tropical forest conservation. In: Collins NM, Thomas JA (eds) The conservation of insects and their habitats. Academic, London, pp 405-422.

Syed, R.A. 1979. Studies on oil palm pollination. Bulletin of Entomology Research, 69:213-224.

Syed, R.A., Law, I.H. and Corley, R.H.V. 1982. Insect pollination of oil palm: introduction, establishment and pollinating efficiency of Elaeidobius kamerunicus in Malaysia. Planter (Malaysia), 58:547-561.

The Telegraph, One third of Europe's butterflies in decline, according to Red List, 2010.

UNEP. 1993. Global biodiversity. UNEP, Nerobi.

vanEngelsdorp, D. et al. 2008. A survey of honey bee colony losses in the U.S., Fall 2007 to Spring 2008. PLoS ONE 3:e4071. doi:DOI:10.1371/ journal.pone.0004071

vanEngelsdorp, D., Underwood, R., Caron, D., Hayes, J. Jr. 2007. An estimate of managed colony losses in the winter of 2006-2007: a report commissioned by the apiary inspectors of America. American Bee Journal, 147:599-603.

Verma, L.R. 1987. Pollination ecology of apple orchids by hymenopterous insects in Matiama Narkanda temperate zone. Final report. Ministry of Environment and Forests, Govt. of India, p 118.

Wilson, E.O. 1988. The current state of biodiversity. In: Wilson EO, Peter FM (eds) Biodiversity. National Academic Press, Washington DC, pp 3-18

Wilson, E.O. 1999. The diversity of life, new editionth edn. W.W. Norton and Company, Inc., New York

Winfree, R. et al. 2007. Effect of human disturbance on bee communities in a forested ecosystem. Conservation of Biology, 21:213-223.

Winfree, R. et al. 2009. A meta-analysis of bees' responses to anthropogenic disturbance. Ecology, 90:2068-2076.

Parmesan, C. 2006. Ecological and evolutionary responses to recent climate change. Annual review of ecological evolution systematic, 37: 637- 669.

Walther, G. R., Pots, E., Convey, P., Menzel, A., Parmesan, C., Beebee, T. J. C. 2002. Ecological responses to recent climatic change. Nature, 416: 389- 395.

Fitter, A.H. and Fitter, R. S. R. 2002. Rapid changes in flowering time in British plants. Science, 296:1689- 1691.

Miller-Rushing, A. J., Katsuki, T., Primack, R. B., Ishii, Y., Lee, S. D. and Higuchi, H. 2007. Impact of global warming on a group of related species and their hybrids: cherry tree (Rosaceae) flowering at Mt. Takao, Japan. American journal of botany, 94:1470- 1478.

Cleland, E. E., Chuine, I., Menzel, A., Mooney, H. A. and Schwartz, M. D. 2007. Shifting plant phenology in response to global change. Trends in Ecology and Evolution, 22(7): 357-365.

Pierre, M. and Cynthia, P. 2013.The Impact of the Nation's Most Widely Used Insecticides on Birds. Neonicotinoid Insecticides and Birds. American Bird Conservancy Publication.

Everts, S. 2008. Honeybee loss- Agriculture: Germany suspends use of clothianidin after the pesticide is linked to honeybee deaths. Chemical Engineering News, 86(21): 10.

Halm, M. P., Rortais, A., Arnold, G., Tasei, J. N. and Rault, S. 2006. New risk assessment approach for systemic insecticides: The case of honey bees and imidacloprid (Gaucho). Environment Science Technology, 40: 2448- 2454.

Iwasa, T., Motoyama, N., Ambrose, J. T. and Roe, R. M. 2004. Mechanism for the differential toxicity of neonicotinoid insecticides in the honey bee, Apis mellifera. Crop Protection, 23: 371- 378.

Jones, A. K., Raymond-Delpech, V., Thany, S. H., Gauthier, M. and Sattelle, D. B. 2006. The nicotinic acetylcholine receptor gene family of the honey bee, Apis mellifera. Genome Research, 16: 1422- 1430.

Tapparo, A., Marton, D., Giorio, C., Zanella, A., Solda, L., Marzaro, M., Vivan, L. and Girolami, V. 2012. "Assessment of the Environmental Exposure of Honeybees to Particulate Matter Containing Neonicotinoid Insecticides Coming from Corn Coated Seeds". Environmental Science & Technology, 46 (5): 2592.

Aufauvre, J., David, G. B., Vidau, C., Fontbonne, R., Roudel, M., Diogon, M., Vigues, B., Belzunces, L. P., Delbac, F. and Blot, N. 2012. Parasite-insecticide interactions: a case study of Nosema ceranae and fipronil synergy on honeybee. Scientific Report, 2: 326; DOI:10.1038/srep00326.

Bernadou, A. Demares, F. Couret-Fauvel, T. Sandoz, J. C. and Gauthier, M. 2009. Effect of fipronil on side-specific antennal tactile learning in the honeybee. Journal of Insect Physiology, 55: 1099- 1106.

Barbara, G. S., Zube, C., Rybak, J., Gauthier, M. and Grunewald, B. 2005. Acetylcholine, GABA and glutamate induce ionic currents in cultured antennal lobe neurons of the honeybee, Apis mellifera. Journal of Comparative Physiology, 191: 823- 836.

Janssen, D., Derst, C., Buckinx, R., Van den Eynden, J., Rigo, J.M. and Van Kerkhove, E. 2007. Dorsal unpaired median neurons of Locusta migratoria express invermectin and fipronil-sensitive glutamate-gated chloride channels. Journal of Neurophysiology, 197: 2642- 2650.

Le Faouder, J., Bichon, E. Brunschwig, P. Landelle, R. Andre, F. andLe Bizec, B. 2007. Transfer assessment of fipronil residues from feed to cow milk. Talanta, 73: 710- 717.

Grant, A. N. 2002. Medicines for sea lice. Pest Management Science, 58 (6): 521- 527.